"Jan Anton provides important insights into sustainable investing. This is an enjoyable and informative book for every investor."

—Elroy Dimson, *Professor of Finance at the Judge Business School, University of Cambridge*

"Sustainable investing has now become a huge and controversial topic but remains surprisingly misunderstood. *Big Money, Better World?* lucidly helps break down many of these misconceptions. More importantly, it persuasively argues that sustainable investors themselves need to evolve from Isaiah Berlin's famous hedgehogs—who know only one big thing—and become more like adaptable foxes that know many little things."

—Robin Wigglesworth, *Author of* Trillions *and* A Fabulous Debt

"Over the past decade, responsible investment has moved from the margins to the mainstream. But with that growth has come a more important question: how much difference is it really making and how can it do more? Not less ambition. Not quieter voices. But better—and bolder—leadership. Because stepping back now won't reduce risk. It will amplify it.

This book engages with that challenge in a way that is both honest and constructive. It challenges some of the simpler narratives that have taken hold in the market and makes a compelling case for a more nuanced, evidence-based approach to responsible investment. What I particularly value is that it doesn't step away from the ambition of using capital to drive change, but instead asks how we do that more effectively, drawing on both research and real-world investment experience.

For those working in responsible investment, this is a timely and important contribution to the next phase of the field."

—Fiona Reynolds, *Former CEO of the UN Principles for Responsible Investment (PRI); President of FAIRR; and Chair of Australian Ethical Super*

"*Big Money, Better World?* is a timely and intellectually rigorous book that brings much needed clarity to the sustainable investing debate. Drawing on the best academic evidence and real-world investment practice, Jan Anton van Zanten cuts through hype and ideology to show when and how capital markets can genuinely contribute to a sustainability transition. This is a rare book that will educate scholars, challenge practitioners, and elevate the public conversation on what sustainable finance can realistically deliver."

—Hao Liang, *President's Chair Professor in Finance, Nanyang Technological University*

"This enjoyable book moves beyond simplistic views of sustainable investing with a refreshingly pragmatic focus on nuance. Because the devil is most often in the details, understanding how capital allocation decisions can drive real change requires looking under the hood and avoiding reductive approaches. As a scientist, I applaud its strong and unwavering commitment to a science-based perspective—and I hope it can once and for all debunk the idea that ESG is synonymous with sustainability, when in fact it is just another financial risk management framework."
—Beatrice Crona, *Professor of Sustainability Science and Science Director of Stockholm Resilience Centre, Stockholm University*

"Van Zanten shows that the problem with sustainable investing is that it over-simplifies a complex problem. Grounded in both research and practice, he provides incredibly pointed and accessible insights into what sustainable investing could be."
—Tima Bansal, *Canada Research Chair in Business Sustainability, Ivey Business School*

"In a field overrun by hedgehogs, Jan Anton van Zanten is the fox we've been waiting for. *Big Money, Better World?* cuts through the noise of sustainable investing with clarity, honesty and a refreshing command of complexity, showing us not what we wish were true, but what the evidence actually says."
—Nina Seega, *Director of the Centre for Sustainable Finance at the Cambridge Institute for Sustainability Leadership*

"*Big Money, Better World?* provides a clear-eyed, evidence-based perspective on sustainable investing. What distinguishes the book is its focus on mechanisms rather than intentions, and its willingness to confront both the potential and the limits of investor influence. Expect no easy answers, but a disciplined framework for thinking about when sustainable investing can plausibly contribute to real-world outcomes—and when claims about impact exceed what the evidence can support."
—Julian Kölbel, *Assistant Professor for Sustainable Finance, University of St. Gallen*

"This book provides a welcome dose of pragmatism to the confusing debates around sustainable investing. Jan Anton makes a compelling case that accounting for not-yet-financial risks and opportunities is simply good investing—and reminds us that perfection shouldn't be the enemy of good. Foxy investors are active investors: they embrace the plurality of the world, know what they own, and know why they own it. A great read for anyone looking to reconnect financial performance with real-world impact."
—Piet Klop, *Director Strategic Relations, and former Head of Responsible Investment at PGGM*

"I love what this book sets out to do. Misconceptions about sustainability still run deep—that's why the book's clarity, nuance, and evidence-based perspective feel so refreshing. This book separates signal from noise with honesty and balance, offering a much needed guide for anyone committed to advancing sustainable investing. I encourage every practitioner to read it and help move our industry toward more informed and impactful action."

—Danielle Brassel, *Head of Responsible Investment at Zurich Insurance Group*

"*Big Money, Better World?* provides deep insights into how sustainable investing should be shaped going forward. It is not just about ESG ratings or dogmatic views, but rather Jan Anton motivates readers to be adaptive, to apply holistic thinking and to continue to ask questions and reassess the underlying motivations and methods surrounding sustainable investing to create real-world impact. A highly recommended read!"

—Ivo Mulder, *Head of UNEP Climate Finance Unit*

"*Big Money, Better World?* highlights a long-overlooked dimension of the role of private sources of finance to support sustainability. It is very timely as it comes in the wake of the launch of the *IPBES Methodological Assessment Report on the Impact and Dependence of Business on Biodiversity* and *Nature's Contributions to People* in February 2026. The book shows how the flow of private capital to 'safe' and supposedly 'risk-free' sustainability initiatives in the global north unfairly impacts lower-middle-income and developing nations. This compounds issues of equity and natural justice that must be addressed for the world to become a safer and more equitable space."

—Chris Gordon, *Founding Director, Institute for Environment and Sanitation Studies, University of Ghana*

"A big part of what is needed when it comes to solving sustainability challenges through investment are more voices calling for practical approaches forward, such as are provided here by Jan Anton van Zanten, an expert whose thinking and work I have long admired. We need more books just like this, more voices of reason, more investment strategy that doesn't leave returns on the table. More of this please."

—Cary Krosinsky, *Lecturer at NYU, Harvard University, and Yale University*

"Jan Anton van Zanten has written a must read on sustainable investing. The book provides an in-depth overview of sustainable investing in companies and countries grounded in theory, practice and empirical evidence. Jan Anton links sustainable investing to SDGs, the guiding strategy used in business. A key insight is that plural approaches towards sustainable investing are needed to achieve sustainable development, which itself is a multi-faceted process."

—Dirk Schoenmaker, *Professor of Banking and Finance at the Rotterdam School of Management, Erasmus University, and author of* Principles of Sustainable Finance

Big Money, Better World?

Jan Anton van Zanten

Big Money, Better World?

Rethinking Sustainable Investing for Real Impact

palgrave
macmillan

Jan Anton van Zanten
Osmosis Investment Management NL
Rotterdam, The Netherlands

ISBN 978-3-032-32441-2 ISBN 978-3-032-32442-9 (eBook)
https://doi.org/10.1007/978-3-032-32442-9

This Palgrave Macmillan imprint is published by the registered company Springer Nature Switzerland AG
The registered company address is: Gewerbestrasse 11, 6330 Cham, Switzerland

For Rosalinde and Wieger
The world is yours

Acknowledgements

Joshua Neoh has been a continuous source of inspiration. He first introduced me to the metaphor of the hedgehog and the fox, which plays a central role in this book, and patiently walked me through numerous philosophical arguments that shaped the thinking behind it.

This book draws heavily on academic research – most notably the work of many outstanding scholars – as well as studies I contributed to myself. I am deeply grateful to my co-authors on the work discussed here: Joop Huij, Dries Laurs, Laurens Swinkels, Bruno Rein, Rikkert Scholten, Lewei He, Harald Lohre, Rob van Tulder, Maria Putintseva, Bhavya Sharma, Malene Christensen, and Anna Vasileva.

I also wish to thank those who reviewed earlier drafts, in whole or in part. Pim van Vliet promised not to pull his punches, and he delivered: his thorough and critical feedback pushed me to substantially rethink the manuscript's focus, and it is stronger for it. Emilio Marti offered sharp, tangible, and constructive comments on major sections. Mathias Larsen's feedback – and our many conversations on sustainability and investing – gave the book a broader perspective. Taeke Wiersma read it with his extensive investment and research experience in mind, helping me sharpen the book's practical orientation. Adriaan Nicolai evaluated it from the perspective of a general reader, reassuring me that the book can speak beyond a finance audience. Rob van Tulder pushed me to reframe the book's central argument – to great effect. Dries Laurs offered a helpful check on the investment implications. Three anonymous reviewers also gave encouraging and useful feedback.

Over the years, I have benefited greatly from the insights of colleagues in the investment industry, currently at Osmosis Investment Management NL and earlier at Robeco. Your support and knowledge run through every chapter.

Thanks, in particular, to Victor Verberk for pushing for more research and innovation in sustainable investment practice.

Thank you also to my editor, Tula Weis, who has played a tremendous role in guiding me through the publishing process and commenting on various aspects of earlier drafts.

I thank my parents for their love and for instilling in me a passion for books, a reverence for nature, and an appreciation of hard work. My sister and grandmother – both named Tallie – brightened my life, each in her own way. I am grateful to Anouk's family for embracing me as one of their own. Above all, I thank Anouk for her unwavering love – and her patience during the writing of this book. Rosalinde, you are my joy. Wieger, you arrived just as Daddy finished the first draft and I am so happy you are here now. This book is dedicated to both of you.

Contents

List of Figures

List of Tables

About the Author

Jan Anton van Zanten writes about sustainable investing from two complementary vantage points: as a practitioner designing sustainable investment strategies and as a researcher studying whether those strategies actually work.

He is Head of Sustainable Investing at Osmosis Investment Management NL and a Visiting Fellow at Erasmus University. Before joining Osmosis, he was Executive Director and SDG Strategist at Robeco, where he worked on aligning equity and fixed income investment portfolios with the ambitions of the Sustainable Development Goals. Earlier in his career, he advised leading financial institutions and multinationals on sustainability strategy at Steward Redqueen and contributed to climate resilience work at the UN Environment Programme.

Jan Anton holds a PhD from Erasmus University, awarded with high honours and recognized with the Erasmus Research Institute of Management's Best Part-time PhD Dissertation Award. He also holds an MPhil from the University of Cambridge and an MSc from Erasmus University. His research has been published in peer-reviewed journals at the nexus of sustainability, finance, and business strategy.

Jan Anton lives in Rotterdam with his partner Anouk and their children Rosalinde and Wieger. He has previously lived in Taipei, Nairobi, and Cambridge for work and study.

1

Introduction

Sustainable Investing at a Crossroads

Big money has a bold ambition: to build a better world. Today, more than 5,000 investors controlling USD 140 trillion have committed to investing sustainably [1]. Their commitments are reshaping markets: an estimated one out of every three dollars is now invested using some form of sustainability integration [2,3]. Sustainability has moved to the top of investors' agendas, and it is increasingly directing how capital flows around the globe.

Two forces explain this rise of sustainable investing.

The first is financial performance. A warming planet, collapsing ecosystems, and widening inequalities are redefining what risk and return mean, causing some companies to win while others lose. By integrating sustainability, investors aim to reduce risk and enhance returns. In short, sustainability can help to protect and grow capital. And this influence will only rise as environmental and social challenges intensify.

Consider Norway's sovereign wealth fund (NBIM), the world's largest investor with USD 1.8 trillion in assets. Its 'universal ownership' strategy means it holds stakes in nearly 9,000 companies – essentially the entire global economy [4]. When systemic threats like climate change emerge, there is nowhere to hide. In 2025, NBIM warned that a 3°C rise in global temperatures could wipe out 19% of its U.S. equity portfolio [5]. When one of the most sophisticated investors sounds the alarm, the message is clear: sustainability is increasingly becoming a financial imperative for many large investors.

The second force driving sustainable investing is impact. Traditionally, investors had two objectives: generating returns and managing risks. Now, a growing set of investors also seek to contribute to a better world through their investment strategies. This elevates sustainability as a standalone investment

J. A. van Zanten, *Big Money, Better World?*,
https://doi.org/10.1007/978-3-032-32442-9_1

objective – explicitly, alongside their traditional financial goals and often reflecting beneficiaries' evolving preferences.

Take PGGM, one of the world's largest pension funds. It recently placed sustainability firmly among its investment objectives – next to returns and risk – arguing that financial results endure only when they also contribute to society, or at the very least do not harm it [6]. This is a quiet revolution. For decades, investors operated in two dimensions: risk and return. Adding sustainability as a third, equally weighted objective fundamentally changes the logic of investing. Sustainability is seen not merely as a financial variable – whether constraint or return driver – but as a goal in its own right.

Although sustainable investing has now become a mainstream ambition, its rise has been fiercely challenged. A backlash has emerged from multiple directions. Some critics deride sustainable investing as 'woke capitalism', arguing it distracts from the core purpose of profit maximization. Others reject it not because it undermines returns, but because it falls short of delivering the systemic change needed to address the planet's most pressing challenges. Tariq Fancy, former chief of sustainable investing at BlackRock, gave this critique its sharpest formulation: he described sustainable investing as a 'dangerous placebo' – something that lets people feel they are making a difference while the actual impact is negligible, and worse, distracts attention from the regulatory and systemic reforms that could produce real change. From this perspective, sustainable investing is seen as tinkering at the margins: just the same old capitalism, now with a green glow.

The result is a deeply divided landscape, involving investors, policymakers, NGOs, and the media. Advocates view sustainable investing as a tool to invest better, or even to help save the planet. Antagonists see it as a distraction from either profit or progress. Amid these polarized views, the conversation is often dominated by self-serving narratives, emotional reasoning, and narrow perspectives. Asset managers overstate their impact, reducing the complexity of sustainable development to a marketing message. Activists dismiss finance altogether, as if capital flows were irrelevant to outcomes. And politicians reduce a complex debate to culture-war fodder, stripping it of the nuance it demands. Each of these responses has one thing in common: a single, oversimplified idea standing in for a messy and interconnected reality. Sustainable investing, consequently, is now plagued by single-idea thinking. But single ideas make poor guides to a complex world.

This book cuts through the noise. It does not claim that sustainable investing will save the world, nor that it is a dangerous distraction. Instead, it offers a rational guide, grounded in the latest scientific research and real-world investment practice, to illuminate the world of sustainable investing. It covers all the big topics: from defining sustainable investing and scrutinizing if investors really invest sustainably to investigating whether it can help build a better world and at what cost. The findings are sometimes surprising. Many investors who claim to invest sustainably do not. And the impact of sustainable investing on the real world is more limited than its advocates typically admit. But nonetheless, these impacts are real – and when scaled, they can make a meaningful difference.

Throughout, it argues that sustainable investing today suffers from 'hedgehog thinking' and needs 'foxy' strategies. The idea comes from *The Hedgehog and the Fox*, a popular essay written by the Oxford philosopher Isaiah Berlin [7]. In it, Berlin draws on a fragment from the ancient Greek poet Archilochus, who said, '*a fox knows many things, but a hedgehog knows one big thing*'. The fox is clever and adaptable, deploying a range of tactics. The hedgehog, by contrast, knows a single but big thing: curling into a ball of spines that no fox can penetrate.

Berlin uses this metaphor to categorize thinkers. Hedgehogs view the world through the lens of a single, overarching idea: a unified vision. Foxes, on the other hand, draw on a wide array of experiences and pursue multiple, sometimes conflicting, goals. For them, the world cannot be boiled down into a single idea. Hedgehogs are single-minded and unwavering; foxes are curious, flexible, and comfortable with complexity.

In sustainable investing, many strategies behave like hedgehogs. They rally around one big idea – portfolio decarbonization, ESG integration, or impact monetization – and treat it as the answer to everything. These approaches can be emotionally compelling, but they often oversimplify the complex reality of sustainable development. The narrow focus makes sustainable investing vulnerable to criticism and leaves practitioners susceptible to disillusionment.

To move forward, we need a foxier approach. Unlike the hedgehog's singular focus, the fox pursues many ideas, embraces complexity, and acknowledges that sustainability is messy and full of trade-offs. Through the lens of the fox, sustainable investing is grounded in sustainability science and rooted in practice, seeking solutions that work, not just those that sound good.

Put simply: sustainable investing fails when it becomes hedgehog-like. A single datapoint or a single rating cannot capture the world's social and environmental complexity. What we need instead is the fox's way of seeing: plural, adaptive, evidence-based. The fox helps us understand sustainability as it actually is, not as we wish it to be. This book shows how sustainable investing can make that shift.

This matters. Sustainable investing holds real potential to improve lives and protect the planet, but potential is not impact. That gap will only close if professionals across the investment ecosystem – asset managers, fiduciaries, regulators, and scholars – understand what sustainable investing truly is, what it can realistically achieve, and how to do it well. That is what this book is for: a rethinking of sustainable investing that is evidence-based, honest about its limits and trade-offs, and focused on real impact. The conclusion it reaches is neither cynical nor naïve: Big money can contribute meaningfully to a better world, but only if it is deployed with the rigour and honesty sustainable development demands.

1.1 This Book

This book takes you through the promises, pitfalls, and possibilities of sustainable investing. It aims to clarify what sustainable investing is, how it works, and where it can contribute to a better world.

I write from two complementary perspectives. As Head of Sustainable Investing, I spend my days designing and implementing strategies that aim to align financial decisions with long-term environmental and social outcomes. As a researcher, I study whether those strategies actually work – and why they sometimes do not. Together, these roles keep me grounded in both evidence and practice and alert to the gap between what sustainable investing promises and what it delivers. That combination runs through every chapter of this book.

Chapter 2 explores what sustainable investing really means – its goals and the pathways to achieve them. But despite its rise, Chapter 3 reveals a paradox: many self-proclaimed sustainable investors do not invest very sustainably. It uncovers this gap and asks why so many investors claiming green credentials invest in so-called 'brown' companies – those causing significant environmental or social harm.

Chapter 4 offers one answer: hedgehog thinking. Dominant approaches to sustainable investing often hinge on one big idea, ignoring the plural nature of sustainable development. This oversimplification blinds investors to complexity and risks missing the point of real-world impact. Chapter 5 introduces a foxy alternative: plural, nuanced approaches that embrace complexity and deliver more effective strategies.

But will sustainable investing actually help solve global challenges? And what trade-offs might investors face in returns? Chapter 6 reviews the evidence for real-world impact, while Chapter 7 examines the financial implications. Together, they show that while sustainable investing will not save the world alone, it can make meaningful contributions. And the evidence is encouraging; this does not require sacrificing returns.

The earlier part of the book focuses on corporate asset classes – equities and credits. Chapter 8 extends the argument to government debt, a vast and underexplored frontier for sustainable investing. Here, the hedgehog view that sustainable investors should favour wealthy nations comes under scrutiny. Instead, the chapter makes the case for channelling capital to low- and middle-income countries, where it can accelerate progress most.

Finally, Chapter 9 concludes by asking the big question: how can sustainable investing scale? It examines the roles of investors, regulators, policymakers, researchers, and individuals – each holding a piece of the puzzle. The conclusion is cautiously optimistic: if these actors work in concert, sustainable investing can move from ambition to impact, and from the margins to the mainstream.

1.2 Actors, Assets, and Issues

Sustainable investing involves a wide range of actors. At its core are asset owners – pension funds, sovereign wealth funds, insurance companies, and university endowments – who hold capital on behalf of beneficiaries, and asset managers, who invest that capital for them. Though their roles and responsibilities differ in important ways, I refer to them collectively as 'investors' throughout this book, unless the distinction matters for the argument at hand. Retail investors – individuals investing their own savings directly – share many of the same underlying principles, though the book's primary focus is on institutional practice, where the scale of capital and the complexity of incentives are greatest.

In terms of what investors can actually own, the options are numerous: listed and private equity, corporate and sovereign bonds, real estate, commodities, hedge funds, and currencies, among others. This book focuses primarily on investments in companies – via listed equity and corporate bonds – and, to a lesser extent, on government bonds. The reason is straightforward: together, these asset classes account for the vast majority of capital deployed worldwide and play the most significant role in shaping both the practice and the impact of sustainable investing.

Sustainable investing must also contend with a daunting range of sustainability issues. Jay-Z once rapped about having 99 problems. The world's sustainability challenges are at least as numerous – and considerably harder to solve. Climate change is widely regarded as the most urgent. The degradation of nature and loss of biodiversity follow close behind, on land, in rivers, and in the seas. Social challenges are equally abundant: inequality and exclusion, peace and justice, access to healthcare, education, and basic needs. Throughout this book, I use the term 'sustainable development' broadly to capture this full spectrum of environmental and social concerns. It serves as an inclusive shorthand for the many dimensions of sustainability that investors must understand and engage with – and as a reminder that no single framework captures them all.

2

Sustainable Investing Is Like a Box of Chocolates

What Is Sustainable Investing, Anyway?

The famous movie *Forrest Gump* opens with a feather floating through the air, eventually landing near Forrest, who is sitting on a bench at a bus stop in Savannah, Georgia. Dressed in a beige suit and holding a box of chocolates on his lap, Forrest begins telling his life story to whoever sits next to him. He opens the box and offers a chocolate to a woman sitting beside him, and says, '*My mama always said, "Life was like a box of chocolates. You never know what you're gonna get."*'

Today, sustainable investing is treated like a box of chocolates. It contains a wide assortment of offerings, each with different ingredients, textures, and flavours. Yet, it is often unclear what lies inside each one, or how they differ from one another.

This ambiguity is reflected in the proliferation of related concepts: socially responsible, ethical, ESG, green, sustainable, impact, and transition investing are all terms used to describe how investors engage with the relationship between capital and sustainability. These labels are frequently used interchangeably. As one group of experts observes, '*One woman's "ethical investing" is another man's "socially responsible investing," and one firm's "responsible investing" is another manager's "sustainable investing."*' They note that '*this embarrassment of semantic richness is perhaps understandable for a rapidly evolving approach, where the final form has yet to be settled*' [8].

Such conceptual overlap inevitably breeds confusion. Investors may simply select whichever flavour sounds most appealing. In the absence of clear and shared definitions of sustainable investing, the landscape becomes one in which hedgehogs reign supreme: simple, compelling narratives overwhelm the plural, interconnected perspectives of the fox.

J. A. van Zanten, *Big Money, Better World?*,
https://doi.org/10.1007/978-3-032-32442-9_2

Before turning to the hedgehog and the fox in greater detail, however, it is worth starting by asking what sustainable investing actually is. To answer that question, it is useful to begin with a brief exploration of the historical development of these overlapping concepts.

2.1 A Brief History of Sustainable Investing

The idea that money should reflect values is centuries old. Religious traditions have long shaped investment norms: Quakers in the seventeenth century refused to profit from weapons and slavery, while Islamic finance has excluded activities such as gambling, pork production, and interest-based lending for generations. In the eighteenth century, John Wesley, founder of Methodism, urged followers to avoid 'sinful trade' and exploitation. By the 1920s, the UK Methodist Church was screening out investments it deemed unethical [9].

Modern approaches to sustainable investing began to take shape in the 1960s, spurred by the civil rights movement and anti-war campaigns that raised awareness of the social implications of investing. The 1970s added an environmental dimension, driven by rising ecological consciousness and reports like those from the Club of Rome that highlighted the ecological limits to unbridled growth [10]. The 1980s saw socially minded investors divest from companies linked to the apartheid regime in South Africa. During this period, the concept of responsible investing began to enter the mainstream.

In the early 2000s, the rise of responsible investing was joined by the popularization of the term ESG, which urged investors to consider environmental, social, and governance factors in their decision-making. A strong push came from the International Finance Corporation's '*Who Cares Wins*' report, which argued that integrating ESG factors into portfolio management would both '*support the growth of sustainable capital flows*' and '*contribute to the sustainable development of societies*' [11]. The global financial crisis further accelerated interest in ESG investing. Having learned the hard way that the financial system might not always serve the public interest – and facing widening macroeconomic inequality – investors increasingly turned to ESG as a way to restore trust and support sustainable development [12].

Despite this early focus on ethics and societal contribution, the 2010s saw the concept of ESG morph into a tool for improving returns. ESG integration increasingly focused on identifying how a company's environmental, social, and governance practices might affect its financial performance. Large investors were attracted to the promised win-win: integrate sustainability and benefit both society and the bottom line. Consequently, this growing emphasis on financial materiality played a key role in mainstreaming sustainability in investment [13].

While generating mass appeal, this also led to the rise of another concept: impact investing. Rather than focusing on the contribution of sustainability to profits, this approach prioritizes positive social and environmental impacts, sometimes even over financial returns. This is the 'dark green' companion to much lighter green ESG approaches.

The term impact investing itself is debated, but consensus has emerged around two main flavours of impact investing: 'buying impact' and 'making impact' [14,15].

The first involves investing in companies that generate a positive impact. The second is more demanding, requiring investors to actively generate impact themselves. This second approach recognizes that simply buying equity or debt from a company may not directly lead to meaningful, sustainable change. More is needed for investors – who are further removed from the real economy – to actively change the world.

Impact investing is not the final concept that has emerged. More recently, yet another one has gained traction: transition finance. This approach focuses on financing the shift from unsustainable business models to sustainable ones, particularly in the context of climate change and biodiversity loss. Rather than investing only in companies that are already sustainable, transition finance targets those that are not yet – but are actively working toward becoming so. It recognizes that the biggest difference might be made by helping 'brown' companies turn 'green'.

These concepts differ in emphasis but share one goal: integrating sustainability into investment decisions. Throughout this book, I will use the term sustainable investing broadly to capture this diversity, while clarifying which type of chocolate we are talking about when it matters for the argument.

2.2 Towards What End?

A useful way to start making sense of sustainable investing is to consider its core objectives. Broadly speaking, there are two: (1) using sustainability insights to enhance financial returns; and (2) aligning financial markets with sustainability goals. These aims are not mutually exclusive. Investors can pursue both – seeking strong performance and positive impact at the same time – though they vary in how much emphasis they place on each.

2.2.1 Improving Financial Returns

For many investors, sustainability is first and foremost a tool to improve financial outcomes. By looking beyond traditional financial metrics and incorporating environmental, social, and governance factors, investors can better assess how well a company is positioned for long-term success.

This can pay off in two ways: protecting against downside risk and unlocking upside potential.

For instance, in 2018 and 2019, two Boeing 737 MAX airplanes crashed within months of each other – one off the coast of Indonesia, the other en route from Addis Ababa to Nairobi – killing 346 people. Investigations revealed a faulty stall-prevention system [16] and a deeper problem: governance failures. Under pressure to cut costs and beat Airbus, Boeing prioritized speed over safety, ignored internal warnings, and sidelined engineering integrity [17]. The result? A global grounding of 737 MAX aircraft, billions in legal settlements, and a 25% share price drop – wiping out USD 62 billion in market value.

BP's Deepwater Horizon spill tells a similar story. In 2010, an offshore rig in the Gulf of Mexico exploded, killing 11 workers and triggering the largest oil spill in U.S. history. In three months, more than 210 million gallons of crude oil – about the same volume as 320 Olympic swimming pools – spilled into the ocean, devastating marine ecosystems, coastal economies, and local communities. The causes of the catastrophe were manifold: ignored warning signs, cut corners on safety, and a culture that put cost savings ahead of environmental risk. The fallout was staggering: BP's stock price plunged 55% in two months, erasing tens of billions in value, while cleanup and fines topped $65 billion [18].

What this means for investors is obvious: analysing sustainability performance is critical for success. Those who consider factors like governance quality, safety culture, and environmental risk may spot red flags that traditional financial analysis misses – helping them avoid costly mistakes. These cases show that sustainability integration is not just a nice-to-have; it is a practical tool for protecting portfolios from catastrophic losses linked to social and environmental mismanagement.

A growing body of research shows that investors increasingly recognize sustainability as a source of financial risk. Companies that perform poorly on climate [19], nature [20], and water stress [21] are no longer seen as neutral – they are viewed as liabilities. When investors finance these laggards, they demand higher returns to offset the added risk, underscoring how deeply sustainability factors now shape investment decisions.

But sustainability is not just about avoiding trouble. The same analytical lens that helps investors identify governance failures and environmental risks can also reveal companies that are well-positioned to thrive in a transitioning world.

Take Enel S.p.A., a multinational energy company headquartered in Rome. Between 2017 and 2024, Enel transformed from a conventional utility into a renewable energy leader – boosting its clean energy capacity by 60%, expanding its portfolio of solar, wind, hydro, and battery storage assets, and cutting its greenhouse gas emissions intensity by 70% – surpassing its own decarbonization targets [22]. It plans to phase out coal by 2027 and reach net-zero by 2040 [23]. This sustainability pivot paid off. Enel's share price nearly doubled, earnings hit EUR 23 billion, and top analysts at Goldman Sachs, UBS, and JP Morgan maintain Buy ratings [24].

2.2.2 Pursuing Sustainability Goals

But investing sustainably just to boost returns is, in a sense, incomplete – like macaroni without the cheese. This matters. If sustainability were only used to improve risk profiles or chase returns, then what would truly set it apart from regular investing? After all, any prudent investor – like your pension fund – should already scrutinize companies like Boeing or BP for risks related to environmental safety, social responsibility, and governance, if only to avoid financial losses. As Alex Edmans, Professor of Finance at London Business School, puts it: '*considering long-term factors isn't ESG investing; it's investing*' [25].

Edmans introduces the concept of '*rational sustainability*' to bridge the financial and societal dimensions of sustainable investing [26]. His argument is pragmatic: sustainability is not an ideological add-on; it is a core part of sound investment practice. Done well, it helps investors identify long-term value drivers – employee well-being, environmental efficiency, governance quality – while contributing to positive societal outcomes. As he writes: '*Good companies create value for both shareholders and society*'. This creates sustainable – meaning long-term – value that, in Edmans' view, should matter to everyone regardless of their professional role, political outlook, or generation. In short, sustainable investing should reflect a dual purpose: better business, better world.

Yet Edmans deliberately avoids prescribing what 'long-term value' means. He leaves open whether rational sustainability is purely about financial value or also about creating societal value – arguing that this is for each investor to decide. What matters, he insists, is transparency: investors should clearly state their objectives, explain how they manage tensions between financial and societal goals, and pursue social values rationally in ways that reinforce long-term financial performance.

Others take a more explicit stance, arguing that sustainable investing must go beyond financial performance and actively pursue social and environmental goals. In this view, it is about aligning capital with sustainability objectives – contributing to solutions for global challenges such as climate change, inequality, and public health.

Such logic led PGGM – the Dutch pension giant – to explicitly add 'impact' to its investment objectives. It explains that this pursuit of sustainability is partly driven by the people who participate in the pension fund. According to PGGM, participants simply want their pension fund to reflect upon the impact their investments are having. Many prefer their money to contribute to a better world and do not want their pension money to be tainted through negative impacts.

Experimental studies among private investors echo this finding: many are willing to pay for more sustainable investments. This diversity of preferences underscores a key point: not all investors are the same. Some prioritize impact as much as, or even more than, financial gain [27].

This impact-driven approach is now embedded in regulation. In 2019, the European Commission introduced the Sustainable Finance Disclosure Regulation (SFDR) to bring clarity and accountability to sustainability claims [28]. Before SFDR, firms could make sweeping statements about 'green' credentials without consistent standards – fuelling greenwashing. SFDR changed that by requiring standardized disclosures and, crucially, defining what counts as a sustainable investment.

Under this definition, a sustainable investment must meet three criteria: make a positive contribution to environmental or social objectives; not cause significant harm to any other environmental or social objectives; must follow good governance practices. In other words, under the European Union's framework, a sustainable investment is one that helps improve the world without simultaneously causing damage.

While the definition leaves room for interpretation – for example, it does not specify exact thresholds for what qualifies as 'positive' or 'harmful' – it firmly anchors sustainable investing in the idea of real-world impact.

Taken together, these ends suggest that sustainable investing must go further than financial performance alone. It is not just about protecting portfolios or chasing higher returns; it is about aligning capital with social and environmental goals to create long-term value for both investors and society. Some investors act from a moral compass, avoiding harmful sectors like tobacco or controversial weapons. Others take a mission-driven approach, aiming to use capital as a lever for change. Whatever the motivation, the essence is clear: sustainable investing is about better business and a better world. But good intentions are only as effective as the strategies used to pursue them – and in sustainable investing, different strategies are at hand.

2.3 Through Which Means?

To better understand sustainable investing, it helps to look not only at its objectives but also at the strategies it uses. Broadly, sustainable investors have two levers of influence: how they allocate capital and how they use their voice as owners. Everything else is a variation on these two themes.

2.3.1 Asset Allocation

In late August 2025, Norway's sovereign wealth fund NBIM sold all its shares in machinery manufacturer Caterpillar, as well as those it held in five Israeli banks. According to the investor's ethics committee, the yellow bulldozers of Caterpillar were being used in the '*unlawful destruction of Palestinian property*', noting that the company '*has not implemented any measures to prevent such use*' [29]. Prior to the divestment, NBIM owned 1.2% of Caterpillar's equity, worth some USD 2.1 billion [30]. Divestment from the Israeli banks was similarly motivated. These banks have provided financial services required for construction activity in the West Bank, which had been '*established in violation of international law*'.

This is a prime example of asset allocation – one of the most popular sustainable investment strategies, and the place where investors 'vote with their feet'. It aims to channel capital toward activities that contribute to sustainable development and away from those that cause harm.

Exclusion – not investing in firms with severe negative impacts, like the NBIM example illustrates – is the strictest way of implementing this strategy. Classic candidates for exclusion are those active in the tobacco, alcohol, gambling, and weapons industries, or those with severe human rights violations. Exclusion sets a baseline – a minimum normative standard of what is considered acceptable. By divesting from these companies, investors send a clear signal that they do not want to fund harm. This strategy is often called 'exit': investors leave harmful companies and redirect capital toward more sustainable ones.

Divestments[1] like these have gained popular traction. At universities, students have been vocal in demanding fossil fuel divestment. The University of Cambridge, for instance, was targeted for investing part of its GBP 3.5 billion endowment in fossil fuel companies. After years of persistent activism from groups like Cambridge Zero Carbon Society, the university pledged to eliminate all fossil fuel investments by 2030 and reach net zero emissions by 2038. The university's Vice-Chancellor Stephen J. Toope called it '*an historic announcement that demonstrates our determination to seek solutions to the climate crisis*', while student leader Ben Margolis described it as '*a landmark decision for which students, staff and the Students' Union have been campaigning for years*'. The fund's Chief Investment Officer Tilly Franklin put the decision in the broader context of rising ecological challenges: '*Climate change, ecological destruction, and biodiversity loss present an urgent existential threat, with severe risks to humankind and all other life on Earth. The Investment Office has responded to those threats by pursuing a strategy that aims to support and encourage the global transition to a carbon-neutral economy*' [31].

Investors can apply exclusions at the whole organizational level – meaning that none of their investment funds can allocate to the excluded company – or at the level of individual investment funds, whereby funds that aim to invest more sustainably will exclude more firms than those with weaker sustainability ambitions.

At the organizational level, there are many investors, especially those in Europe, who publish lists of excluded companies, along with the reasons. NBIM excludes around 200 companies, for issues ranging from coal and nuclear energy to human rights violations, controversial weapons, and environmental damage. Dutch pension fund giants ABP and PGGM, respectively, exclude around 300 and 500 companies – adding criteria like Arctic oil drilling or involvement in tar sands. Danish pension fund Danica excludes over a thousand companies.

[1] Exclusion is often also referred to as divestment. They are not synonymous, however. Divestment is simply the selling of a company's stocks or bonds. Naturally, companies that are added to an investor's exclusion list need to be divested (sold) if they are held in portfolios. Here, I simply use exclusion and divestment interchangeably.

At the level of individual investment funds, there is much more diversity. Some funds brand themselves as being 'dark green'. They apply much stricter sustainability screening than 'light green' or regular investment funds, thus excluding a larger portion of the investment universe. By developing diverse portfolios with different levels of stringency in sustainability screening, investors aim to cater to the diverging interests in sustainability. Some clients will be satisfied with a light green fund, while others will seek a solution that has a much stronger focus on pursuing sustainability goals.

Of course, exclusion comes with trade-offs. It limits opportunities for generating financial performance. If a divested company later delivers strong financial returns, those gains are missed – potentially frustrating portfolio managers and raising questions from asset owners. That is why exclusion is often seen as a rigorous measure that is reserved for the worst offenders: those whose impacts are too severe to ignore.

But asset allocation is not just about exiting companies you do not like. It is also about seeking to finance those who are part of the solution. This can take the form of best-in-class strategies that invest in the most sustainable firms in each sector. Or it can form part of a thematic investment strategy that only invests in companies providing solutions to a particular sustainability challenge, like the shift towards renewable energy, better healthcare, or water and sanitation. And then there is tilting: rather than directly excluding companies, investors may allocate less financing to companies with a weak sustainability profile and tilt more capital towards those with stronger profiles.

Yet when you invest in an entire company, or country for that matter, you finance its entire operations – from the good to the potentially bad and ugly. How might you, as an investor, more directly invest in sustainability progress?

Specialized debt instruments have been developed that make sure that investors really finance sustainability projects. The most famous is the green bond [32], though there are many others [33], including social bonds, development bonds, and even rhino bonds.[2] These are debt instruments with the key requirement that the funds raised must go toward projects with environmental or social benefits. These bonds help issuers finance sustainability improvements and give investors a credible way to allocate capital directly to solutions.

[2] In 2022, the Global Environment Facility and World Bank launched the five-year, USD 150 million Wildlife Conservation Bond – better known as the "Rhino Bond." The proceeds are earmarked to fund rhino conservation in South Africa. At maturity, investors recover their principal and may earn success payouts – up to USD 13.76 million – tied to black rhino population growth.

For example, a chemical company might issue a green bond to finance the development of clean solutions. Investors owning the bond then know their money is tied to green activities, rather than the brown projects the firm may also undertake.

Beyond bonds whose proceeds are earmarked for sustainable activities, there are sustainability-linked bonds. These work differently. Companies issuing such bonds set targets for achieving company-wide sustainability goals. If they meet them – meaning they are becoming more sustainable – they get a discount on their borrowing costs. They benefit by paying less; investors do by financing sustainability improvements.

Thus, through asset allocation, investors direct money to whole companies – or to specific sustainability-oriented instruments like green bonds – that support positive sustainability outcomes. But asset allocation is ultimately a relatively passive strategy. Active ownership offers something different: rather than simply directing capital, it allows investors to use their influence to actively shape company behaviour in line with environmental and social goals.

2.3.2 Active Ownership

Whereas asset allocation is often referred to as 'exit', active ownership is widely known as 'voice'. It has two dimensions: proxy voting and engagement.

Proxy voting is a strong tool that investors can wield. Owners of a company's equity can cast their votes on important issues at shareholder meetings without needing to be physically present. In sustainable investing, proxy voting is often used to support resolutions related to climate action, diversity, human rights, executive pay, and other matters. Investors can vote for or against proposals, board members, or corporate policies, signalling their expectations for more responsible business practices. When coordinated across large shareholder coalitions, proxy voting can drive meaningful change, such as pushing companies to adopt net-zero targets or improve transparency on social issues. It is a way for investors to turn ownership into stewardship.

A widely discussed example concerns activist investor Engine No. 1. In 2021, it led a campaign to elect three new board members at the oil firm ExxonMobil [34]. Engine No.1 argued that Exxon's leadership failed to adequately respond to climate risks, evidenced by the company's lack of a credible strategy for transitioning away from fossil fuels. This, according to Engine No. 1, threatened long-term value creation.

Engine No. 1 nominated four independent directors who had experience in the energy transition. While the investor only owned 0.02% of Exxon's shares, it started a campaign that gained traction with some of the largest institutional investors in the world, including the big three: BlackRock, Vanguard, and State Street. Ultimately, three of the four nominees were elected, unseating sitting board members.

While the campaign sent a powerful message to boardrooms around the world, its actual effect on Exxon's climate performance is less clear. In the years since, Exxon has continued to expand oil production and has escalated its lobbying against climate regulation – a reminder that changing a board is not the same as changing a business model [35].

A more recent illustration comes from BP's annual general meeting in April 2026 [36]. Unlike Engine No. 1's activist campaign, this was active ownership by mainstream investors. More than half of BP's shareholders voted against the company's attempt to retire its climate disclosure obligations that it had made to investors in earlier years. A further majority opposed a move to virtual-only AGMs, which some viewed as an attempt to keep sustainability-related discussions at bay, while the newly appointed chair was elected with just 82% support, well below the near-unanimous approval typically afforded to board chairs. The revolt was backed not by a small activist fund but by major institutional investors including Legal & General, several UK pension funds, and proxy advisers Glass Lewis and ISS. The message was clear: shareholders were not willing to let BP quietly walk back transparency commitments it had previously made. Where Engine No. 1 showed what a determined minority can achieve, BP 2026 showed what happens when the institutional mainstream speaks with a collective voice.

Further to filing and voting on resolutions, investors can engage in a dialogue with the companies they invest in. Investors typically contact the company's investor relations team to initiate dialogue. This approach aims to influence a company's behaviour by raising concerns, asking questions, and encouraging improvements. Engagement can be done through private meetings, collaborative investor initiatives, or public letters. Whereas proxy voting is episodic and formal, engagement is ongoing and relational.

Firms can benefit from active ownership as well: they can strengthen their relations with investors, learn about best practices that investors see take hold in other companies they own, and improve their practices.

A good illustration of a collaborative engagement initiative is Climate Action 100+. Launched in 2017, CA100+ brings together over 700 institutional investors, representing more than USD 68 trillion in assets, to engage with companies on three topics: improving climate governance, reducing emissions, and enhancing climate-related financial disclosures. The collective approach ensures that there is a lot of financial firepower behind the engagement requests. It also ensures that investors target companies in a consistent manner, improving efficiency and reducing costs for the companies that are being targeted – which is also in the interest of investors.

A success story is the engagement with RWE AG, a German utility company historically reliant on coal. Through sustained dialogue and pressure from Climate Action 100+ signatories, RWE committed to phasing out coal and aligned its emissions reduction targets with the 1.5°C pathway, validated by the Science Based Targets initiative (SBTi). This marked a significant shift in strategy and demonstrated how collaborative investor engagement can support meaningful change even in carbon-intensive industries [37].

Active ownership, thereby, overcomes a drawback associated with asset allocation: once you divest from a company, you lose your seat at the table. You forfeit the chance to influence and improve the firm's behaviour. And if you sell your stake, someone else will own it. Chances are that the buyer might care less about sustainability than you do. Active ownership keeps you at the negotiating table, allowing you to use your voice and call for change. Nevertheless, progress can be slow, especially if there is no majority support among shareholders for the desired changes [38], and since the power of active ownership has its limit – after all, how do you turn a tobacco company into a health-promoting enterprise? – sustainable investors ultimately resort to exit, redirecting capital away from companies where engagement has failed.

For example, when divesting from the six Israeli companies, NBIM explained that: '*Before deciding to exclude a company, Norges Bank shall consider whether other measures, including active ownership, may be better suited. The Board's assessment is that it is not appropriate to use other measures in these cases*' [39].

Combined, asset allocation and active ownership offer a complementary set of tools for sustainable investing. Through asset allocation, investors can align the money they manage with the sustainability outcomes they want to see in the world, while active ownership offers them a voice to call proactively for positive change.

2.4 Defining What Sustainable Investing Is and Is Not

Having explored the ends and means of sustainable investing, it is worth pinning down more precisely what it is – and what it is not. Definitions in this field are contested. Some frame sustainable investing primarily around financial risk management – integrating ESG factors to protect returns. Others define it by its instruments, focusing on specific products like green bonds or impact funds. Still others anchor it in investor intent, distinguishing those who seek measurable change from those who merely want to avoid reputational harm. Each of these framings captures something real, but each also leaves something important out. The definition I adopt throughout this book is deliberately broader in one respect and narrower in another: broader, because it encompasses both asset allocation and active ownership as legitimate pathways; narrower, because it requires a real commitment to sustainable development as an objective – not merely as a financial lens.

Based on the discussion above, I adopt the following working definition:

> '*Sustainable investing involves asset allocation and active ownership activities that aim to support sustainable development and create long-term value*'.

This definition aligns with how academic experts typically frame sustainable investing [40]. I favour it because it places the emphasis where it belongs: on advancing sustainability goals through the investment practices of asset allocation and active ownership – and because it draws a clear line between sustainable investing and prudent investing. An investor who integrates sustainability solely to manage financial risk is doing something valuable, but it is not, by this definition, sustainable investing. That distinction will matter throughout the chapters that follow.

This definition rests on two key components.

The first is *sustainable development*. While the term may sound aspirational or vague, it is central to the purpose of sustainable investing. At its core, sustainable development means meeting the needs of people today without compromising the ability of future generations to meet theirs. It implies safeguarding ecosystems, promoting human well-being, and enabling inclusive economic progress [41].

In practice, this means sustainable investors must engage with a wide range of sustainability issues – from climate change and biodiversity loss to poverty, inequality, conflict, and lack of access to basic needs. These challenges span both environmental and social domains. Addressing them requires a broad lens.

Through this first component of the definition, sustainable investors aim to support a diversity of sustainable development objectives. I use the term support deliberately, rather than stronger language like create or deliver. This reflects a realistic appraisal of investor influence: while investors can contribute meaningfully to advancing sustainable development, the causal channels are complex, and their ultimate impact is often indirect – as we will explore in detail in Chapter 7.

The second component is creating *long-term value*, which encompasses two types of value.

The first is straightforward: generating financial returns for clients. Investing, by definition, involves deploying capital with the expectation of return. In a transitioning world, sustainability is increasingly relevant to long-term financial performance, making it a necessary condition for any credible investment strategy. But for sustainable investors, financial performance alone is never sufficient.

The second type of value is broader. It includes benefits derived from aligning portfolios with sustainable development – such as knowing your investments reflect your values, supporting improved social and environmental outcomes, and enhancing reputation or stakeholder trust. These dimensions will be explored further in later chapters.

The notion of long-term signals that sustainable investors adopt extended time horizons recognizing that money managed today shapes outcomes for future generations. This is especially relevant for large institutional investors – such as sovereign wealth funds and pension funds – whose mandates span decades.

This dual purpose matters. An investor who integrates sustainability solely to improve returns or manage financial risk is not, in this book's definition, investing sustainably. That may be prudent investing, but it is not sustainable investing. The distinction is important: sustainable investing requires a genuine commitment to supporting positive environmental and social outcomes, not merely using sustainability as a financial lens.

With this definition in place, it is worth clarifying what sustainable investing is not – and how it relates to the many concepts in the sustainable investing box.

In practice, ESG integration pursued purely to avoid financially material risks and capture profitable opportunities – without also aiming for better environmental and social outcomes – does not qualify as sustainable investing. ESG integration could support positive progress in individual investment cases, but it frequently does not, as Chapter 4 will show.

Some concepts are part of sustainable investing but go further. Impact investing is one example. It is stricter: impact investors not only aim to support sustainable development but also seek to intentionally generate measurable social and environmental benefits, using a theory of change that identifies how their actions ultimately shape outcomes – and how these can be both measured and managed [42]. With these more demanding criteria, impact investing is most frequently applied in asset classes where it can be best implemented, such as private equity and debt, real assets, project finance, and green and social bonds – and less so in public equity and debt markets. This makes impact investing one of the more demanding varieties within the broader box of sustainable investing.

Other concepts, like transition investing, offer important nuance within the broader field. Initially, sustainable investing focused on financing firms that already supported sustainability goals, or at least did not cause harm. Transition investing broadened this focus to include firms that may not be sustainable today but are implementing credible actions to improve their impact in the future. With this logic, transition investing aims to support the journey, helping brown companies become greener. This too falls within the scope of sustainable investing.

Sustainable investing is also not marketing and coalition building. Many investors feature sustainability in their marketing campaigns and join coalitions that rally around a sustainability objective. Both can be relevant and help build an environment for positive change. But it is not enough: sustainable investing is about market strategies – asset allocation and active ownership – that can be adopted to advance sustainable development.[3]

2.5 Concluding Remarks: Exit and Voice for a Better World

Sustainable investing is not one thing. It is an umbrella term that encompasses a wide range of activities at the intersection of finance and sustainability, and people hold different conceptions of what it is, what it should achieve, and how it should be practised. Umbrella terms can be useful: they allow conversations to move forward without getting bogged down in overly technical definitions. But some level of shared understanding is essential. Without it, we risk talking past each other: using the same words while meaning very different things.

This chapter offered a solid understanding of sustainable investing without being overly prescriptive. Sustainable investing aims to support sustainable development and create long-term value, using two main levers: asset allocation and active ownership. One keeps you in the game to push for change;

[3] As chapter 7 will show, investors can adopt non-market strategies to promote sustainable development, such as 'field building'. These are important and deserve attention. Yet, since they do not involve investment practice, I am reluctant to label such strategies as 'sustainable investing'.

the other lets you vote with your feet when change proves elusive. Through asset allocation – or exit – investors can steer capital toward sustainability leaders and away from laggards. Through active ownership – or voice – they can use their power as shareholders to call for improvements in corporate behaviour. Together, these strategies form a complementary set of tools for aligning investment portfolios with sustainable development goals. Sustainable investing may resemble a box of chocolates, but it is still one thing: using the tools of exit and voice to shape a more sustainable world.

But here is the uncomfortable question that the rest of this book keeps returning to: does any of this actually work? Investors have committed trillions to sustainable strategies. They have signed pledges, launched green funds, and published ambitious targets. Yet what investors say and what they do are often very different things – and the gap between promise and practice is larger and more consequential than the industry typically admits. The chapters that follow examine that gap honestly: what sustainable investors actually do with their capital, why the tools they rely on so often fall short, and what a more rigorous, evidence-based approach would look like. The stakes are high enough that getting this right matters.

3

Everything That Glitters Is Not Gold

Do Sustainable Investors Invest Sustainably?

In 2020, DWS – the asset management arm of Deutsche Bank – hired Desiree Fixler as Group Sustainability Officer. Soon after, she uncovered a big gap between DWS's glossy sustainability promises and its actual investment practices [43]. According to Fixler, the firm's communications were exaggerated and misleading. When she raised these concerns internally, conflict ensued, and she ultimately became a whistleblower. After a lengthy investigation, DWS was fined USD 27 million for greenwashing and quietly scaled back its sustainability claims [44].

This high-profile case illustrates a timeless truth: not everything that glitters is gold. In sustainable investing, impressive claims do not always translate into meaningful action.

And today, the investment industry is awash with impressive claims. Visit the website of any fund manager or asset owner, and you will likely find bold statements linking global sustainability challenges to the investor's financial expertise. For example, State Street promises to support investors '*in achieving their sustainability-related goals and invest with confidence*' [45]. Vanguard suggests its ESG funds '*can make an impact beyond your portfolio*' [46]. PIMCO emphasizes '*the transition to a resilient, lower-carbon energy system*' [47]. And Amundi frames its mission as part of a broader movement to '*efficiently allocate capital to ensure sustainability of the ecosystem within which we are collectively operating*' [48].

But how could we identify which investors are sustainable beyond their marketing claims?

One way is to look at whether or not they have signed up to various sustainable investing initiatives. There are many initiatives that investors can

J. A. van Zanten, *Big Money, Better World?*,
https://doi.org/10.1007/978-3-032-32442-9_3

commit to, which induce them to invest more responsibly. And many investors have made a public commitment to them, causing them to signal their green credentials.

Without a doubt, the most prominent of these is the UN-backed Principles for Responsible Investment (PRI). Today, the PRI boasts over 5,000 signatories committed to building a sustainable financial system that '*benefits the environment and society*' [49]. Other initiatives include the Institutional Investor Group on Climate Change (IIGCC) [50], Net Zero alliances under GFANZ [51], and Climate Action 100+, who rally hundreds of investors behind decarbonization goals. And then you have Finance for Biodiversity [52] and Nature Action 100 [53], which add momentum on biodiversity and nature.

Together, these initiatives represent trillions of dollars, with numerous investors portraying themselves as having sustainable investment ambitions. This should inspire optimism: if the world's largest investors are backing sustainable change, progress seems possible. Yet, optimism comes with caveats. Of course, ambition alone does not guarantee impact. The real test lies in whether these glittering commitments translate into the gold of more sustainable investment practices.

This brings us to several key questions: do investors who claim to lead on sustainable investing invest sustainably? What types of factors determine whether an investor will invest more or less sustainably? And beyond the investors themselves, are sustainable investment funds truly allocating capital to sustainability leaders? Let us tackle these questions step by step.

3.1 The Owners of (Un)Sustainable Companies

Traditionally, the investment industry operates as a rational, evidence-based practice. Investors pursue returns and manage risk, both of which are easily quantifiable. At any moment, you can see how much money a particular investment is yielding and how well an investor is doing compared to their benchmark. This allows for a clear assessment of whether they are outperforming or underperforming the broader set of available investment opportunities.

But working in sustainable investing can feel less rational. While it is easy to count how many sustainability initiatives and principles and investor endorse, it is much harder to measure how sustainable their portfolio really is. Although sustainability is increasingly treated as a distinct objective for investors to pursue, there are few widely accepted yardsticks to assess how well investors are doing from a sustainability perspective. Investors share many stories, but navigating this space can feel like moving through a fog: it is difficult to see who is really investing sustainably and whose portfolios are rather unsustainable.

The rapid rise of sustainability initiatives in the investment industry inspired me to look for evidence. The question I wanted to shed light on was simple but critical: do investors who sign sustainability pledges actually invest more sustainably? My interest was specifically in asset allocation: do investors who have committed to sustainable investing initiatives hold fewer shares in unsustainable companies – those causing environmental or social harm – and more in sustainable ones that offer solutions?

Together with Bruno Rein, who was interning with me at the time, we began by creating three groups of companies. The first group consisted of 286 *unsustainable* companies that cause social or environmental harm, with a combined market value of USD 4.1 trillion. The second group included 351 *sustainable* companies that provide solutions for improving human or planetary wellbeing, worth USD 7.1 trillion. The third group contained 477 *neutral* companies – those with no significant impacts on sustainable development – valued at USD 4.9 trillion [54].

Because sustainability performance is not a black-or-white (or in this case brown-or-green) decision, we focused only on those companies that were widely recognized as either sustainable leaders or unsustainable laggards. In other words, we included companies in the sustainable and unsustainable buckets only when there was broad consensus about their impact – either positive or negative.[1]

With these groups established, we collected a lot of data about them over a period of 10 years. Most importantly, we identified who the 50 largest investors are in each of the companies, year by year. This gave us a detailed view of who owns (un)sustainable companies.

From there, we gathered more information about these investors: their geographic location, what type of investors they are (e.g., pension fund, insurance company, hedge fund, etc.), and whether they had committed to major sustainable investing initiatives like the PRI, IIGCC, and Net Zero alliances.

[1] To sort companies into these groups, we combined several independent indicators rather than trusting any single sustainability rating - a safeguard against the subjectivity involved in deciding whether firms are (un)sustainable. A company counted as unsustainable only if it had a negative SDG score and met at least one further test: wide exclusion by large asset owners (for example, for producing controversial weapons, tobacco, or thermal coal), very high greenhouse-gas intensity, or significant harm under the EU taxonomy. A company counted as sustainable only if it had a positive SDG score and either featured in dedicated sustainable funds or earned most of its revenue from EU-taxonomy-aligned activities.

The big picture view revealed that just five investors – BlackRock, Capital Group, Fidelity, State Street, and Vanguard – jointly own around 20% of the total market value of all (un)sustainable companies. Yet this concentration of ownership matters enormously for sustainable development. Earlier research, by a team of sustainability scientists and finance researchers, called these firms 'Financial Giants' and showed that they have considerable influence on the world's climate through their outsized ownership of companies operating in the Brazilian Amazon rainforest and the boreal forests in Russia and Canada [55].

In addition to examining the individual owners of (un)sustainable firms, there were some intriguing regional differences. Investors from Europe, East Asia, and the United Kingdom allocated more of their assets to sustainable companies relative to those that are unsustainable.[2] In contrast, investors based in the United States, South and Southeast Asia, and Oceania allocated more financing to unsustainable companies.

But perhaps most surprising were the findings about the actual investment behaviour of those who publicly commit to sustainable investing.

We found that signing up to sustainability initiatives does not significantly increase an investor's ownership of sustainable companies, nor does it reduce their ownership of unsustainable ones. In general, the investment behaviour of signatories resembled that of non-signatories. We had expected that investors who endorsed sustainability initiatives would hold fewer unsustainable companies and more sustainable ones, relative to their holdings in neutral firms. But contrary to this expectation, signatories behaved no differently from non-signatories: both groups held more sustainable and more unsustainable companies than neutral ones. This implies that they did not actively steer assets away from negative impacts or towards positive contributions.

Even when comparing investors that committed to all three initiatives – PRI, IIGCC, and Net Zero alliances – with those who had not committed to any, the differences were minimal. This comparison between the most enthusiastic endorsers to the quiet non-committers painted a similar picture: neither group could be said to invest much more sustainably than the other. From the tests we conducted, only PRI members seemed to invest slightly less in unsustainable companies than non-members – although it should be noted that this effect was very small and not particularly meaningful.

[2]The groups of sustainable and unsustainable companies differ in size – both in terms of how many companies are included and what their combined market value is. To see whether investors have a higher ownership of (un)sustainable companies, we measured their ownership divided by the market capitalization of all companies in each sample. This normalizes the groups, allowing for a fair comparison.

We also compared investors who joined the PRI in its early years (2006) with those who signed on much later (2018 or beyond). Early movers were found to invest less in unsustainable companies relative to neutral ones, while late joiners invested more in unsustainable companies. Ownership of sustainable firms was similar across both groups.

This suggests that some investors – those who have engaged with sustainability for a longer time – actively try to steer away from harmful investments. Others, by contrast, appear to engage with sustainability more superficially: embracing the language of sustainability without significantly altering their investment behaviour.

The results complement studies that have investigated other effects of signing up to sustainability initiatives. For example, investors that have committed to the PRI are found to build portfolios with lower sustainability risks and to further reduce these risks over time – but only if they are based outside of the United States [56]. U.S.-based investors, by contrast, were found to treat the PRI more opportunistically. They often sign but fail to integrate ESG meaningfully. This indicates that many investors may treat sustainability initiatives primarily as tools for managing financial risks and opportunities – rather than as mechanisms for steering capital towards solving sustainability challenges.

Overall, these findings indicate that sustainable investing initiatives do not appear to be very effective in directing capital towards more sustainable outcomes. Many studies before ours have raised suspicions that such initiatives may promote greenwashing [56,57], and our results echo those concerns.

So, should we dismiss these initiatives as window dressing? Not yet. There is another lever: active ownership.

In Chapter 2, we saw that asset allocation is only one of the main strategies available to sustainable investors. Another is active ownership: using your power as an investor to vote at companies' annual meetings and engaging with them to promote sustainable outcomes. Many sustainability initiatives focus on stewardship, encouraging investors to push for change at the level of the companies that they invest in.

Following this strategy, investors actually need to own unsustainable companies. Doing so gives them a seat at the table to start dialogues and cast votes. Under this logic, owning unsustainable companies is not hypocrisy but strategy.

Unfortunately, this channel can probably be ruled out as well. One study, which analyzed a decade of voting data using more than 20 million observations, found that signatories of the PRI do not vote more often in favour of environmental and social issues than non-signatories [58].

Another study went further, examining what happens after investors sign the PRI, using data on U.S.-based investors [59]. It found that after committing to the PRI, investors did not increase their support for pro-sustainability shareholder proposals. Likewise, there was no evidence that signatories improved their engagement practices. The authors concluded that '*a reasonable reader may perceive our findings as consistent with PRI funds' greenwashing*'.

3.2 What Shapes Sustainable Investment?

So far, the picture looks mixed: sustainability pledges rarely shift capital flows. But that is not the whole story. Are there other forces that shape whether investors finance sustainability leaders or laggards? A growing body of research suggests that certain investor traits may make them more or less likely to invest sustainably.

3.2.1 Normative Pressure

One force is normative pressure. Just as the behaviour of individuals is guided by norms, investors in the markets also face normative pressure. A well-known norm in finance discourages financing 'sin stocks', companies involved in activities like tobacco, gambling, alcohol, and controversial weapons. More recently, high-carbon firms have joined the list [60], reflecting a growing norm against financing global warming.

But not all investors are subject to the same level of normative pressure. A landmark paper published in 2009, entitled 'The Price of Sin', hypothesized that investors with a more public profile face greater normative pressure than those operating outside the public eye [61]. Pension funds, universities, sovereign wealth funds, banks, and insurance companies are closely watched by society. In contrast, other investors like hedge funds, mutual funds, and individuals face less public scrutiny [61, 62]. These less-visible investors often act as natural arbitrageurs in the market: when one investor sells a (sin) stock, another must buy it.

The paper's findings support this hypothesis. Investors under greater normative pressure tend to invest less in sin stocks. Specifically, these investors hold approximately 18% lower ownership in sin stocks than in comparable firms. The conclusion is important: normative pressure appears to be a mechanism that can steer capital away from unsustainable companies – thereby encouraging more sustainable investment action.

This effect is not just based on a single year. The analysis spans the period from 1980 to 2006. Has anything changed in the two decades since?

Not much. Normative pressure still seems to guide investor behaviour. In our own study, where we looked at the owners of (un)sustainable companies [54], we also looked at the influence of normative pressure. We found that investors facing greater normative pressure tend to invest less in unsustainable companies, relative to their ownership of neutral companies. However, they do not invest more in sustainable firms. In other words, investors under public scrutiny – like pension funds or sovereign wealth funds – actively avoid unsustainable companies but do not necessarily target sustainability leaders.

3.2.2 Legal Systems

Characteristics of an investor's home country can also influence sustainable investment behaviour [54]. One of these is the country's legal system, whereby a broad distinction can be made between common law and civil law systems.

In common law countries, like the U.S. or the U.K., fiduciary duty is framed narrowly: maximize shareholder wealth. Sustainability, if addressed at all, is often approached reactively, with a focus on avoiding liability rather than proactively solving problems. The view is short-term: if sustainability does not boost profits in the near term, investors may be seen as breaching their fiduciary duty to maximize returns. Sustainability is often reactive, seeking to avoid liability, not to pursue impact. Unsurprisingly, investors here hold more unsustainable firms – but also more sustainable ones – than neutral ones, reflecting a profit-first mindset.

By contrast, civil law countries, such as Germany, France, or Scandinavian nations, recognize broader stakeholder interests, encompassing employees, communities, and the environment. This legal backdrop encourages the avoidance of harm but also the pursuit of positive impact. Profits remain important, but do not always trump social and environmental responsibilities. Investors in these systems consistently hold fewer unsustainable firms and slightly more sustainable ones. Thus, investors from these countries tend to reveal a more consistent pattern of sustainable investing.

3.2.3 National Commitment to Sustainability

A final trait that influences sustainable investing is a country's approach to sustainable development [54].

Some countries are strongly committed to promoting sustainable development, while others are less so. This can be witnessed by their national plans, budgets, investments, and monitoring systems. Analysis by the Sustainable Development Solutions Network – a global think tank linked to the UN – revealed that countries like Denmark, Japan, Austria, Mexico, Nigeria, and Colombia are highly committed, while United States, Israel, Brazil, Bolivia, Russia, and Vietnam are examples of nations with very low commitment [63].

We found that investors from countries whose governments are strongly committed to sustainable development tended to invest much more sustainably. They own lower stakes in unsustainable firms and higher ones in sustainable firms, compared to investors from countries with less dedication to sustainability.

Taken together, these traits suggest that external forces play a powerful role in shaping sustainable investment behaviour. Informal social norms, formal legal systems, and countries' commitment to sustainability are key factors influencing how investors allocate capital – often more so than an investor's own proclaimed sustainability ambitions.

3.3 Green Funds, Brown Companies

All of these findings suggest that there is no clear-cut archetype of a sustainable investor. Many investors may genuinely support sustainable development objectives at an organizational level, yet still manage portfolios that do not consistently reflect those ambitions – for instance, because their clients are not demanding it. As a result, an investor might manage a few 'dark green' portfolios alongside a broader set of conventional investment funds that are largely agnostic about sustainability.

This possibility could feel somewhat schizophrenic: an investor might manage a few funds that invest in companies offering sustainability solutions, while simultaneously allocating most of their capital to companies causing the problems.

This reality raises an important question: how sustainable are so-called sustainable investment funds?

Journalists from Follow the Money and Investec explored this question in their '*Great Green Investment Investigation*', published in 2022 [64].

They examined funds classified as 'Article 9' under Europe's Sustainable Finance Disclosure Regulation (SFDR). These funds are considered 'dark green'. According to the regulation, they should only invest in companies that contribute to environmental or social objectives, without causing significant harm, and while adhering to good governance standards.

At the time of their investigation, there were 1,141 such funds, of which they analyzed 838. They evaluated the companies that these funds invest in and assessed their sustainability performance, particularly from a climate change perspective.

Despite their green claims, the journalists found that almost half of these funds also invest in fossil fuels and aviation. For example, 20funds invest in Delta Air Lines, 10 in Equinor (one of Europe's largest natural gas companies), and others in Shell, Total Energies, and Easyjet. These are clear cases of companies you would not expect in a highly sustainable investment portfolio – though some other holdings are more debatable.

Take NextEra, for instance. This American energy company received more investment from 'dark green' funds than any other company involved in fossil fuels. While it operates many wind farms, NextEra also generates half of its power from gas and coal. The regulation does not clearly define what qualifies as sustainable investing, leaving significant room for interpretation and for investors to decide what is green enough'.

In response to these findings, the European Securities and Markets Authority (ESMA), the EU's financial markets regulator and supervisor, told Investico and Follow the Money that it is '*rather difficult to argue*' that fossil investments do not cause social or environmental damage. Paul Smeets, Professor of Sustainable Finance at the University of Amsterdam, noted that many of these 'dark green' funds charge higher fees. This presents a double risk to clients: they pay more for a fund marketed as sustainable, which may not be green at all [65].

More extensive academic research has since corroborated these findings. One study examined whether light (Article 8) and dark (Article 9) green portfolios differ meaningfully from funds without a sustainability objective, using analytical techniques borrowed from biology and ecology to assess the similarity of fund holdings [66]. Covering 6,888 equity funds invested across 24,549 assets, the researchers found that sustainability-oriented funds invest in largely similar companies as conventional ones: for every sustainable fund, there is a non-sustainable fund with very similar holdings. While dark green funds show some tendency to avoid sin stocks, there is little evidence that either light or dark green funds invest in companies that regular funds do not.

Various factors could explain why self-proclaimed sustainable funds do not always invest sustainably. A cynic might see this as greenwashing, and in some cases, that perception is justified. There are certainly instances where investors have presented a rosier picture to the outside world than what is reflected in their portfolios. The DWS case mentioned in the opening of this chapter is a telling example.

But not all cases stem from deliberate misrepresentation. The Dutch investment firm Actiam, for example, had 12 'dark green' funds in the database studied by the journalists from Follow the Money and Investico. Only two of these funds did not make any investments in brown companies. The remaining 10 made a total of 279 investments in fossil fuel-related firms. Actiam's head of sustainability was reportedly shocked by the findings, responding: '*if what you say is true, then we will immediately sell these positions*', and adding that the regulation is '*crap*', since it does not clearly define what qualifies as a 'sustainable investment' [65].

An academic study of 'Article 9' funds – the EU's designation for the most sustainable investment products – supports the view that investors may not be entirely sure what constitutes a sustainable investment. Examining over 1,000 of these dark green funds, researchers found that only 60% deliberately targeted companies promoting environmental or social impact. The remaining 40% incorporated sustainability factors primarily to enhance financial performance, not to drive positive change. In other words, a substantial share of funds appears to miss the impact they claim to pursue, a finding the authors attribute to '*ambiguity around how to determine what constitutes sustainable investment*' [67].

While you may rightly be amazed that some so-called sustainable investors appear uncertain about whether fossil fuels belong in sustainable funds, these examples highlight the urgent need to explore how investors assess the sustainability performance of companies.

3.4 Concluding Remarks: Diagnosing Before Prescribing

This chapter underscored a wise adage: everything that glitters is not gold. This is particularly apt in the sustainable investment industry. Strong branding and slick marketing prevail. Many investors publicly commit to sustainability initiatives, signalling their intent to invest responsibly. But when you examine their actual strategies, they often resemble those of investors who make no such commitments. Both groups tend to own similar shares in sustainable and unsustainable companies – and show little difference in how actively they engage with them.

This green mirage is also visible at the fund level. The only way to determine whether a fund is truly sustainable is to look under the hood – at its holdings and stewardship practices. When you do, you will often find that many 'green' funds contain 'brown' companies. As a result, the boundaries between green investing and greenwashing seem to blur.

Different reasons explain why self-proclaimed sustainable investors and green funds do not always invest very sustainably. One is deliberate greenwashing: if appearing sustainable wins business without the effort of actually investing that way, some will seize the premium. Another reason is less cynical: investors may simply not know which companies are truly sustainable. Their motives may be sincere, but if they cannot tell the difference, they fail at the most fundamental task of sustainable investing.

A useful analogy comes from healthcare. In the Netflix series *New Amsterdam*, Dr. Max Goodwin rushes to treat a patient showing signs of a heart attack. The symptoms, however, stem from a rare neurological condition. The initial treatment – standard for cardiac care – worsens the patient's condition. Only after a deeper diagnostic process does the team uncover the real issue and adjust the intervention. The lesson is clear: good intentions are not enough. Before a doctor can treat a patient, they must first understand what is wrong. Diagnosis is the foundation for effective care.

Sustainable investing faces a similar challenge. Before allocating capital to a company or engaging with it, investors need a clear picture of how that company performs across sustainability dimensions. Otherwise, the investment may fail to support meaningful progress – or worse, inadvertently finance harm.

That is why sustainability screening is so critical. Just as doctors rely on blood tests, scans, and patient histories, investors use data, ratings, and analytical frameworks to evaluate corporate sustainability performance. These tools help distinguish between companies that lead on sustainability and those that lag behind.

The good news is that investors have access to an array of sustainability analytics. The bad news is that there is almost too much: tools, metrics, and ratings have proliferated to the point of overwhelming those trying to use them. While this abundance may seem promising, many tools obscure more than they reveal. With so many approaches competing for attention, it becomes difficult to discern which methods offer insight and which simply add noise. The problem, as the next two chapters will argue, is not a lack of data – it is a flaw in how that data are being used. Chapter 4 examines why the dominant screening methods behave like hedgehogs, rallying around a single idea at the expense of sustainability's true complexity. Chapter 5 then asks what a foxier alternative would look like in practice.

4

When Hedgehogs Rule Sustainable Investing

Do Singular Sustainability Screening Methods Make Sense?

My friend Joshua Neoh and I met in 2015 at Cambridge, as we both matriculated into Gonville and Caius College. At the time a Lecturer in legal theory at the Australian National University, Josh had come to Cambridge to write his PhD dissertation. His research explored the relationship between law, love, and freedom – a thoroughly philosophic topic that seemed to me both intriguing and amusing.

During college dinners, we would frequently discuss moral philosophy, guided by Josh's Socratic style. There would be no bounds to the conversations at the long candlelit tables in the college's dining hall: from life in the state of nature (would it really be nasty, brutish, and short?), to the origins of human rights (are you born with them, or does someone bestow them upon you?), and to the many variations of the trolley problem (is it worse to steer a trolley towards people, or hurl people towards a trolley?). We even debated the ethics of animal welfare, including why microwaving kittens is morally abhorrent – a memorable test of moral intuition.

At some point, Josh introduced me to Isaiah Berlin's metaphor of the hedgehog and the fox and their different approaches to moral values. Hedgehogs interpret the world through a single overarching idea – a unifying system to which all else is subordinate. Foxes, by contrast, pursue many ends that may be unrelated or even contradictory, without insisting that everything fit into one coherent framework.

The distinction has enjoyed a long afterlife in popular culture, at quasi-intellectual dinner parties, and across academic disciplines, because it invites fun speculation: is a given thinker – or even someone at the table – a hedgehog or a fox?

J. A. van Zanten, *Big Money, Better World?*,
https://doi.org/10.1007/978-3-032-32442-9_4

Economics provides vivid examples. Karl Marx, with his all-encompassing vision of class struggle and a materialist conception of history, is a classic hedgehog. John Maynard Keynes, by contrast, is the quintessential fox. Famously described as '*having more ideas before lunch than most people have in a lifetime*', Keynes ranged across disciplines, shifting his views as evidence changed, and drawing together disparate, sometimes contradictory ideas in order to solve real-world problems [68].

When Josh first shared this metaphor with me, I was researching how companies were responding to the newly launched Sustainable Development Goals (SDGs). That same philosophical question – whether there are many equally valid and potentially conflicting values, or one fundamental value under which all others can be unified – became surprisingly relevant. I began asking myself whether the 17 SDGs represent a plural set of sustainability values – health, education, climate, poverty, and gender equality – each important in its own right, and not reducible to the others?

Or could they be collapsed into one overarching value, such as human development, or economic growth, or planetary stability, with the individual goals treated merely as means to that single end?

A simple example illustrates the tension. Consider two sustainability values: urban development and environmental preservation. A fox would view both as valid, distinct, and potentially in conflict. Cities may expand at the expense of ecosystems, and protecting ecosystems may constrain urban development. Both values matter, and no single rational measure tells us which should dominate. A hedgehog, however, would identify one overarching value – say, economic growth – and treat both urban development and environmental protection as instruments serving that ultimate end: growing the economy.

How is this metaphor relevant to sustainable investing?

I believe it provides a powerful lens for understanding the wide-ranging diversity of sustainability screening methods available to investors. The countless datapoints, metrics, and ratings that investors encounter differ radically in their purpose, scope, and methodology. Some approaches are tailor-made for hedgehogs; others are decidedly fox-like.

This chapter surveys the hedgehog's approaches to sustainability screening. These distinctions help reveal how investors simplify sustainability – and where blind spots inevitably emerge.

I argue that in sustainable investing, we can distinguish among three types of hedgehogs. The first is the *Specialist Hedgehog*, who fixates on a single sustainability issue – climate change, biodiversity, or human rights – while sidelining the rest. The second is the *Filtering Hedgehog*, who considers many issues but views them all through a single overarching lens, such as financial materiality. Finally, there is the *Accountant Hedgehog*, who also looks across multiple topics but reduces them to a single monetary value. Each approach simplifies complexity, but each does so in very different ways.

Let us examine these types in turn, to assess whether their tactics make sense for constructing sustainable investment portfolios – and to imagine how the fox might raise an eyebrow at each of them.

4.1 The Specialist Hedgehog

One prominent hedgehog in sustainable investing is the one that specializes. This *specialist hedgehog* targets one single sustainability issue, ignoring all others. Prominent topics in which sustainable investors specialize are greenhouse gas emissions, water and waste, and green revenues.

4.1.1 Greenhouse Gas Emissions

The current affairs magazine, The Economist, declared that sustainable investing '*should be boiled down to one simple measure: emissions*' [69]. The logic, it argued, was straightforward: climate change is '*by far the most significant danger*', so investors should focus on identifying which companies emit the most – and which are reducing emissions most rapidly.

In its view, sustainable investing covers such a sprawling set of social, environmental, and governance themes that there are simply too many targets to aim at, resulting in '*less chance of bullseye-ing any of them*'. With so many disparate objectives, The Economist suggested that investors should narrow their attention to environmental issues alone. Yet even that, it claimed, was still too broad, since 'the environment' encompasses biodiversity, water scarcity, pollution, and countless other topics.

The magazine's conclusion was therefore to focus solely on climate change, given the imminent risks posed by '*emissions … generated by carbon-belching industries*'. The imperative, it insisted, was to screen companies primarily on their greenhouse-gas emissions in order to pinpoint the biggest polluters and reward those cutting emissions most aggressively.

Here, The Economist epitomizes the specialist hedgehog's approach to sustainable investing. It does not deny the relevance of other sustainability objectives, but it elevates greenhouse emissions as the holy grail of sustainable investing. This hedgehog tactic is not contrarian. Climate change is widely – and rightly – seen as one of the greatest challenges of our time.

Beyond its urgency, focusing solely on climate change has another advantage: there is a lot of data that investors can use. Companies typically report their greenhouse gas emissions across three scopes. Scope 1 covers emissions that are directly generated by the company. Scope 2 captures indirect emissions from a company's consumption of purchased electricity, heat, or steam. Scope 3 encompasses all other indirect emissions in a company's supply chain (upstream) and the use of its products (downstream).

The widespread availability of emissions data has enabled investors to build climate-conscious portfolios. By financing low-emission firms and avoiding high emitters, investors aim to align their portfolios with the goals of curbing global warming.

Indeed, emission reduction is one of the most popular objectives in sustainable investing. Morningstar analysis shows that, as of September 2024, roughly USD 168 billion was invested in passive strategies aiming to reduce emissions relative to climate benchmarks [70]. Another USD 50 billion was allocated to active strategies with similar goals. This makes emissions reduction a hugely popular objective in sustainable investing today.

Yet despite its popularity, focusing solely on emissions can mislead. Emissions data, while important, tells an incomplete story about a company's climate performance.

One challenge is Scope 3. While Scopes 1 and 2 are relatively well measured, Scope 3 remains elusive. Methodologies are still evolving, and many companies do not report these emissions at all. This has serious implications. A fossil fuel company may appear cleaner than an electric utility when only Scopes 1 and 2 are considered – because the emissions from burning its fuels (Scope 3) are ignored. Conversely, companies with high Scope 1 and 2 emissions may actually be climate heroes, their footprint a byproduct of manufacturing the very solutions the energy transition depends on.

This has a big impact on investment practice. Focusing narrowly on emissions data can lead investors to underweight companies offering climate solutions and overweight those contributing to the problem. In short, it would make it likely that they would finance exactly the wrong companies, making emissions data seem more meaningful than it is.

Another challenge is interpretation. Emissions are usually reported in metric tons – but how can investors know whether a company's footprint aligns with global climate goals? To address this difficulty, some rating agencies have developed more evolved metrics. One of the most advanced is the temperature score, which estimates how a company's emissions trajectory aligns with the 'Paris Agreement's goal of limiting warming to 1.5°C or 2°C.

Temperature scores act like a climate compass – pointing investors toward companies whose emissions trajectories align with global goals. Every company emits carbon, but not all are equally misaligned. Temperature scores assign a value – measured in degrees Celsius – to each company. A firm might be a '1.1°C company', meaning its emissions trajectory aligns with a 1.1°C rise in global temperatures. Another might be on a '3.4°C pathway', indicating it is far from Paris-aligned. Portfolio-level scores can be averaged and 'benchmarked', allowing managers to showcase their alignment with climate goals.

MSCI's Implied Temperature Rise metric is a pioneering example. It assigns sector-specific carbon budgets, projects future emissions to 2050, and compares these projections to the budget. Companies with projected emissions below the budget receive lower temperature scores; those that overshoot get higher ones.

In theory, this is a powerful tool. It links individual companies to global climate outcomes, acting like a searchlight that highlights where climate action should focus. It is also practical: despite its complex construction, the interpretation is intuitive – the further a company is from the 1.5°C target, the worse its climate impact.

But temperature scores have flaws. Like raw emissions data, they do not reward companies for producing climate solutions – such as renewable energy, battery technology, or electric vehicles. This leads to surprising results. Take Tesla: despite its central role in the EV transition, MSCI rated it as aligned with a 2.63°C pathway – far from Paris-aligned.

Moreover, the precision of these scores has been questioned.

Jakob Thomä, founder of the 2 Degrees Investing Initiative, has been among the most vocal critics of this pseudo-precision – arguing that claiming a company is on a 2.63°C rather than a 2.62°C pathway implies a level of accuracy that is simply not consistent with the uncertainty inherent in the underlying data, the forward-looking projections, or climate science itself. The concern is not merely technical: precision that outruns the evidence gives investors false confidence and obscures the very uncertainty they most need to understand. Shortly after such criticisms gained traction, MSCI changed its scores to one decimal place in June 2022 [71], a welcome adjustment, although this does not alleviate the concern that they might imply too much accuracy.

While climate change rightly dominates the conversation, it's only one part of a company's environmental footprint. Other dimensions – such as water use, waste generation, and social factors like gender equality or employee safety – have gained prominence in sustainable investing.

4.1.2 Water and Waste

Let us focus briefly on environmental impacts. Data on water and waste are becoming more available, driven by regulation and stakeholder pressure. For companies that do not disclose, data providers offer estimates – though accuracy remains a challenge. Incomplete or inconsistent reporting can mislead, but the growing availability of this data allows investors to steer portfolios toward the issues they prioritize [72].

Compared to greenhouse gas emissions, screening companies on water and waste introduces two notable complexities.

First, geographic context matters. Two breweries may report identical water usage, yet their impacts could differ dramatically if one operates in rainy Scotland while the other draws from scarce supplies in arid South Africa. Unlike greenhouse gas emissions – which affect the climate globally – water stress is inherently local. Some companies acknowledge this nuance. For example, Coca-Cola HBC reports that in 2024 it used 18 million cubic metres of water, half sourced from areas facing water risk [73]. But few firms offer this level of disclosure, limiting the ability to gather meaningful, comparable data for a wide range of companies.

This brings us to the second complexity: comparability is limited. A company producing hazardous waste has a very different footprint than one generating non-toxic waste – even if volumes are the same. Emissions are typically reported in standardized CO_2 equivalents; water and waste are not.

And finally, impacts outside of the firm's factory gates are typically ignored in footprint data. Beer brewer Heineken explains that in 2024 it used 3.1 hectolitres (hl) of water to brew 1 hl of beer [74]. This number dropped 11% since 2018, thus showing a positive trend. Nevertheless, it is only a part of the story. Most of a beer brewer's water impact does not take place in the brewery but in the supply chain – where vast amounts of water are used to grow agricultural inputs such as barley and hops.

Over time, these measurement challenges may be resolved. But even then, metrics like water, waste, or greenhouse gas emissions only assess how sustainable a company's operations are. They often ignore the impact of their products, as we have seen with the wind turbine manufacturer that appears unsustainable due to emissions-intensive production, despite enabling the energy transition.

Ultimately, screening based on operational metrics – whether emissions, water, or waste – offers only a partial view. To truly assess sustainability, investors must also consider the impact of a company's products and its role in driving change. So, how can investors understand if companies sell sustainability solutions?

4.1.3 Green Revenues

This is where green revenues come in. Green revenues are a metric showing how much of a company's sales are coming from products or services that help address environmental challenges. Most often, green revenues focus on selling solutions for climate change, although other categories are gaining traction, including those linked to the circular economy, pollution prevention, ecosystem restoration, and sustainable use of water and marine resources.

Green revenues are a metric that is of great benefit in overcoming the challenges with footprint data, such as greenhouse gas emissions. As we saw, a wind turbine manufacturer has high emissions, which are generated during the production process. Green revenues complement the picture by measuring the value of such a company's sales. If the company is fully dedicated to selling wind turbines, 100% of its revenues would be green. This way, green revenues help to identify which companies are selling solutions for tackling sustainability challenges – going far beyond operational impacts.

Yet how do you determine which products are green – and can thus be captured by green revenues – and which are not?

A prominent framework for measuring green revenues is the EU Taxonomy. According to the European Commission, achieving the EU's climate and energy targets – and the broader objectives of the Green Deal – requires '*directing investments towards sustainable projects and activities*' [75]. To do so, a shared language and a clear definition of what counts as 'sustainable' are essential.

That is precisely the purpose of the Taxonomy: it classifies environmentally sustainable economic activities that investors can finance. It sets out technical screening criteria to determine when a company's activities qualify. In short, an activity must contribute to one or more of six environmental objectives, do no significant harm to any of them, and meet minimum social safeguards.

The level of detail is striking. Hydropower, for instance, is considered a climate mitigation solution – but only under strict conditions: the plant must be run-of-river without an artificial reservoir, have a power density above 5 W/m^2, and produce life-cycle emissions below 100 gCO_2e/kWh. Aluminium manufacturing can qualify too, but only if direct emissions stay below 1,484 tCO_2e per tonne, indirect emissions remain under 100 gCO_2e/kWh, and electricity use does not exceed 15.5 MWh per tonne [76].

This specificity makes the Taxonomy rigorous – but also complex. Evaluating companies against these criteria is time-consuming, prompting many investors to rely on specialized data providers, who have built a lucrative business around selling green revenue data.

Despite its complexity, the Taxonomy offers clear advantages. It is politically endorsed and developed with input from sustainability researchers – reducing, though not eliminating, the subjectivity involved in assessing sustainability performance.

But are green revenues the solution to better sustainability screening?

No. Despite their real and valid contributions, they are not the ultimate end. Green revenues add everything that is 'green' together in a single number. This hedgehog view would insist that 'green is green', and that there is no need to differentiate between various shades of green. However, this does not match reality. Not all green products are equally meaningful.

For example, both solar energy and hydropower contribute to climate mitigation. Yet researchers explain that solar has far greater potential to reduce emissions [77]. Lumping all that is green together in one metric is hard to reconcile with sustainability science.

Moreover, in addition to differentiating between different shades of green revenues, it is important to evaluate green revenues in relation to brown ones. Two companies may have an equal share of green revenues – say 50%. Yet if one firm derives another 50% of revenues from thermal coal, while the second gets its remaining share from a more neutral product like data technology, then the latter is clearly better for the world than the former. Negative impact must also be accounted for.

This challenge raises the broader question: does it serve investors well to address a single sustainability topic, or are they better off taking a foxier strategy that addresses multiple topics?

4.2 How the Specialist Hedgehog Overlooks Sustainability's Plurality

At first glance, the specialist hedgehog seems to be in an enviable position. By focusing on a singular topic – like greenhouse gases or green revenues – this hedgehog enjoys a straightforward sustainable investment strategy. The data are relatively easy to interpret, judgments require limited subjectivity, and the process avoids the feeling of sustainability screening being a 'black box'. The logic is simple: the higher the footprint, the worse the company; the higher the green revenues, the better. And when you combine footprint and revenue data, you account for both the company's operations and the impact of its products.

But those with a foxier mindset would raise a deeper concern – one that goes beyond methodology. Companies often have multiple, and sometimes conflicting, impacts on different sustainability goals.

What should we make of a company that manufactures solar panels – advancing climate goals – but mistreats its workers? Or a firm that generates significant waste but also delivers promising innovations in healthcare?

In such cases, the specialist hedgehog might dismiss this complexity by insisting that the chosen objective is most paramount, much like how The Economist argued that greenhouse gas emissions should be the ultimate focus of sustainable investing. This does not deny the existence of other problems, but places one urgent challenge at the heart of sustainable development. Solve this one, and the rest can wait.

This reductionist view runs counter to how sustainable development actually works. Sustainability challenges rarely exist in isolation. Climate change, biodiversity loss, water scarcity, pollution, and inequality are deeply interconnected, often reinforcing one another. Rising temperatures can worsen water shortages, which in turn could affect food production, public health, and even conflict.

This is why sustainability experts increasingly call for a 'nexus' or integrated approach: one that recognizes how environmental and social systems are intertwined and seeks solutions that address multiple goals at once [78].

This interconnected reality means that assessing a company's sustainability performance requires more than looking at isolated metrics. A firm might excel in one area – say, reducing emissions – while falling short in others, such as labour rights or water use. Understanding sustainability, therefore, demands a holistic view that weighs trade-offs and synergies across different goals. It is not just about doing less harm in one domain or pursuing solutions in another. It is about contributing positively across the board: minimizing trade-offs and chasing synergies.

This is not just a philosophical concern: it affects real-world portfolios in practice.

The S&P 500 Biodiversity Index [79], for instance, holds a sizeable position in GE Aerospace – a company whose operations significantly contribute to global warming and therefore, indirectly, to biodiversity loss. Its military products are also used in conflicts around the world, raising difficult questions about their broader social impacts. Likewise, the MSCI World Climate Paris Aligned Index [80] includes Amazon and Meta. Neither company contributes to addressing climate change in any direct or obvious way. At the same time, both are widely associated with negative impacts on other sustainability dimensions: Amazon has been frequently criticized for its labour practices, while Meta has been accused of promoting harmful content, influencing elections, and contributing to mental-health issues [81].[1] Although the precise rationale for their inclusion is not always transparent, the presence of such companies invites critical reflection on whether mainstream sustainable investing meaningfully advances sustainable development.

These examples underscore a fundamental concern: sustainability strategies that screen for a single theme risk financing companies that perform well on that one issue while causing significant harm on others. A screening approach focused narrowly on individual topics fails to reflect the integrated, plural nature of sustainable development. It overlooks the inherent tensions and trade-offs between sustainability goals.

The answer to this challenge is, in principle, straightforward. To align with sustainability science, investors must assess companies across multiple sustainability goals and synthesize those insights into an integrated view. Only then can we arrive at an aggregate sustainability assessment that captures the full picture rather than a single slice of it. Hedgehogs, as we will see next, have devised one approach that comes close to this ambition: ESG ratings.

---

[1] For purists: Paris-Aligned Benchmarks (PAB) and Climate Transition Benchmarks (CTB) are EU-regulated climate benchmarks designed to support the transition to a net-zero economy. They require benchmark providers to reduce the carbon intensity of the index relative to the investable universe – by at least 50% for PAB and 30% for CTB – and then follow an annual 7% decarbonization trajectory. Both benchmarks also exclude companies involved in certain high-carbon or unethical activities.

What makes these benchmarks distinctive is that sustainability screening is applied primarily at the portfolio level, not at the company level. Emissions reductions are largely achieved by excluding high emitters, rather than by requiring companies within the benchmark to decarbonize. This is why a common critique is that PAB and CTB can look climate-aligned on paper yet deliver limited real-world impact. Moreover, as climate-only frameworks, they ignore broader sustainability dimensions beyond climate.

4.3 The Filtering Hedgehog

The filtering hedgehog goes beyond simply screening companies on a single topic – like greenhouse gas emissions – and evaluates many themes. The prime tactic it uses is ESG screening. At first sight, ESG ratings seem foxy. They evaluate companies on their performance on numerous environmental, social, and governance (ESG) indicators.

Environmental factors could, for example, focus on greenhouse gas emissions, toxic spills, waste, or water. Social indicators might screen companies on child labour, employee development, or broader labour practices. And the governance dimension typically investigates remuneration, separation of the company's CEO and chairperson, or collective bargaining.

Having screened companies on all of these factors, an ESG rating is calculated that gives an overall impression of the performance of a company. This calculation would typically reflect the divergent performance of a company on different topics.

This seems right up the fox's alley. But looks can deceive. ESG ratings cover many issues, yet filter them through a single lens, creating a distinctly hedgehog approach. Before unpacking how that works, let us step back and examine the outsized role ESG ratings now play in financial markets and the peculiar quirks that come with them.

4.3.1 ESG Ratings Move Markets

The earliest ESG ratings trace their history to the 1980s. Those early ESG ratings had an explicit ethical focus, catering to a niche market of religious and norms-based investors. Today, they rank among the most dominant metrics used in the sustainable investing industry.

A survey among a large group of global investors revealed that in 2020, integrating ESG into the investment process was by far the most important approach to sustainable investing, reaching USD 25.2 trillion in assets under management. This equalled 71% of all sustainably managed assets at that time [2]. For comparison, this value is larger than the combined GDP of the European Union and Japan.

In fact, ESG ratings have become so popular that they move markets. Investment funds that select companies with better ESG ratings experience sizable inflows of money. Those targeting companies with poor ESG ratings tend to see outflows. An intriguing study by Samuel Hartzmark and Abigail Sussman provided causal evidence for this relation [82].

They capitalized on a natural experiment to see how important ESG ratings really are in shaping investment strategies. In 2016, the data provider Morningstar first published sustainability ratings for more than 20,000 mutual funds – together accounting for over USD 8 trillion in assets. All of these investment funds were given a total rating based on the companies they invested in. The top 10% was given five globes (high sustainability) and the worst 10% gained one globe (low sustainability) – with 80% of funds in the middle receiving two, three, or four globes.

This totally changed sustainable investing: for the first time, the sustainability performance of investors' funds shifted from something difficult to understand to something that was very clearly displayed and advertised by one of the industry's leading research organizations.

What happened next was striking. The researchers showed that in the nine months before the rating was published, the funds received similar flows. But when the sustainability ratings were published, the investors started to change their behaviour. They allocated more money to the best-ranked funds with five stars and less money to the worst funds with one star – whereby funds in the middle with two, three, or four globes were not really affected.

Did this really matter, or was it simply a statistical effect?

To find out, they quantified the value of the flows. Funds with the best sustainability rating experienced substantial inflows of money, approximating 4% of fund size in the course of nine months. In contrast, poorly performing funds saw outflows of money of roughly 6% of the size of the fund. In total, the researchers estimated that in this period after publication, between 12 and 15 billion USD were taken out of funds with poor sustainability performance, and between 24 and 32 billion USD were put into the top-performing funds. In short, this study showed that ESG integration helps funds attract more money, benefitting the funds' managers[2], thereby creating more demand for ESG ratings and propagating the trend of using such sustainability metrics in the investment process.

This study was among the first to provide evidence of how sustainability information influences the behaviour of investors. More studies have followed, showing similar effects.

In another – somewhat amusing – study, researchers looked at what happened when Sustainalytics, a leading provider of company ESG ratings, inverted the scale of its ESG rating [83]. Originally, higher scores indicated better, and lower scores worse, ESG performance. Yet in September 2019,

[2] Fund managers benefit from inflows as their earnings often partly depend on the size of the fund.

Sustainalytics decided to inverse this rating, with lower scores signalling good performance and higher scores indicating bad performance.

What happened next was rather unfortunate. Investors treated the rating-scale change as if it signalled real 'downgrades', and traded accordingly – selling names that now carried lower scores and buying those that appeared 'upgraded'. This misinterpretation of a cosmetic scale inversion drove divestments and reallocations, increased trading costs, and hurt performance for investors and their beneficiaries. It also demonstrated how strongly ESG ratings can influence investment decisions even when underlying fundamentals are unchanged.

So, ESG ratings are moving markets. Investment funds that finance companies with better ESG performance attract money, while those with poor ESG ratings lose out.

4.3.2 Confusion and Bias

Despite their prominence, ESG ratings have drawn various criticisms. One concern is their lack of consistency. Different providers often give wildly different scores for the same company. Many companies with good ratings from one agency receive a low or medium assessment by another [84].

Tesla is a prime example. Some agencies reward their role in accelerating the electric vehicle transition. Others penalize it for the emissions of its factories or its governance controversies and labour issues [85]. As a result, Tesla can be an ESG darling in one index and a pariah in another. Elon Musk famously called ESG 'a scam' after S&P dropped Tesla from its ESG index.

If some rating agencies give a company like Tesla a good score while others rate it poorly, where does that leave investors? The answer is: confused.

A landmark study – aptly titled *Aggregate Confusion* [86] – analyzed ratings from six major providers and found striking divergence, even when broken down into environmental, social, and governance components. Environmental scores aligned the most, governance the least, with social factors somewhere in between. The paper identified three main sources of divergence: scope (agencies assess different topics), measurement (they use different data for the same topic), and weighting (they assign different importance to the same issue).

Measurement differences alone account for more than half the variation.[3] As the authors conclude, ESG divergence does not make measurement

[3] Their results showed that measurement divergence was the dominant driver, accounting for 56% of the disagreement, followed by scope (38%) and weighting (6%). To quantify these effects, the team analyzed 709 indicators across the six ratings, categorizing them into a unified taxonomy.

futile, but it does make it messy. Investors need to understand what lies beneath the scores before relying on them. Or, as one commentator quipped, '*A company can be an ESG saint according to one rating agency and a complete environmental terrorist according to the other. It's Schrödinger's sustainability – you are simultaneously green and brown until someone opens the ratings box*' [87].

In short, ESG ratings can vary wildly depending on who is doing the judging. But regardless of who is judging, a second critique has emerged: biases in ESG ratings. Three stand out: size, location, and disclosure.

First, *size* matters. Larger companies tend to receive higher ESG ratings than smaller ones, both within and across industries. One explanation is that larger firms have more resources to invest in environmental and social initiatives. But the same logic provides a counterargument: size also brings power and the ability to lobby for unsustainable policies, pressure suppliers, or market harmful products. 'Big oil', 'big pharma', and 'big tobacco' are shorthand for exactly these dynamics.

Second, *location* influences. Companies from wealthier countries often receive higher ratings, partly because stricter regulations make it harder to pollute or exploit workers. Yet this logic is far from airtight. Many multinationals shift negative impacts to poorer regions with weaker oversight. Meanwhile, innovators in emerging markets – like Kenya's M-Pesa, which revolutionized mobile payments and lifted millions out of poverty [88] – may score poorly despite delivering real-world benefits.

Third, *disclosure* is rewarded. Firms that publish more sustainability data tend to score higher, as rating agencies often penalize non-disclosure. But disclosure is not performance. Greenwashing is rampant: KLM's 'Fly Responsibly' campaign was ruled misleading [89]; Shein was fined for deceptive claims [90]; and Coca-Cola claimed that its bottles are fully recyclable even though they are not [91].

The result? ESG ratings frequently favour large firms, rich-country firms, and those that excel at sustainability storytelling, although there is no obvious reason why these might really be more sustainable. One analyst described it to me with some frustration: these biases create so much noise that ESG ratings become nearly useless for determining how sustainable companies actually are. This leads to an intriguing question: do ESG ratings really measure how well firms contribute to social wellbeing and safeguarding nature?

4.3.3 Missing Impact

I frequently give lectures to university students and professionals in the finance industry about sustainable investing. During these sessions, I like to do a simple experiment to test how well students' perceptions of a company's sustainability performance match the ESG ratings that investors use.

I show the logos of four companies and ask students to raise their hands if they believe the firm belongs in a sustainable investment portfolio.

The first company is British American Tobacco (BAT). The name speaks for itself: BAT is one of the world's leading cigarette manufacturers, holding around 19% of the global cigarette market outside China [92]. It grows tobacco, manufactures cigarettes, and distributes brands like Lucky Strike, Camel, Kent, and Pall Mall in 91 countries.

Given the well-known health impacts of smoking, it is unsurprising that nobody raises their hand.

Next is Glencore, a Swiss-based natural resources giant. It has global mining operations that provide metals and minerals, like copper, nickel, and zinc. Such materials are useful for industrialization, infrastructure development, and the green transition. It also produces energy, including from oil and coal, for steelmaking.

What is remarkable about Glencore is its legal track record. It has faced legal scrutiny and accusations of corruption, bribery, fraud, and market manipulation around the world. In 2022, Glencore pleaded guilty in the United States to charges of corruption and bribery spanning over a decade. It admitted to making corrupt payments to officials in Nigeria, Cameroon, Ivory Coast, Equatorial Guinea, Brazil, Venezuela, and the Democratic Republic of the Congo (DRC). Glencore also pleaded guilty to manipulating two benchmark price assessments for fuel oil products. Speaking of these cases, US Attorneys said that "*the scope of this criminal bribery scheme is staggering*" and "*Glencore's market price manipulation threatened not just financial harm, but undermined participants' faith in the commodities markets' fair and efficient function that we all rely on*" [93]. In these cases alone, the Swiss firm paid over USD 1.1 billion in fines.

After hearing this, students understandably keep their hands close to the ground: this does not sound like a company that a sustainable investor should pursue.

The third company I show is SunPower Corporation. This California-headquartered company specializes in solar panels and battery storage solutions, mainly for residential homeowners. Having been founded in 1985, it has been one of the early pioneers in the fast-moving renewable energy sector. The company's history has been volatile from a financial perspective. While it saw large investments – and had, for example, TotalEnergies, the French oil firm, purchase a controlling stake in the firm for USD 1.37 billion in 2011 – in 2024 it faced bankruptcy, a restart, but also turned profitable and cash flow positive [94]. This long history, combined with an estimated 5% market share, makes SunPower a relevant player in the U.S. renewable energy space.

Most people associate solar energy with sustainability. Rocky financial performance cannot deny the contribution SunPower has played in enabling homeowners to obtain access to renewable energy. Hands are raised all around to indicate that this indeed could be a company that a sustainable investor might want to investigate.

The final company is Evoqua Water Technologies, which produces water and wastewater treatment solutions. Its clients are companies that use water in their manufacturing process – such as those active in manufacturing, healthcare, pharmaceuticals, biotech, and power. During the production processes of such companies, contaminants can pollute the water that is being used. Evoqua's water treatment solutions are helpful because they ensure that when this water is returned – to nature or municipal water systems – the pollutants are captured and stay behind. It was founded in the U.S. in 2013. In 2023, it was acquired by Xylem, a global water technology company [95].

Delivering technology that helps ensure that water stays clean naturally sounds very sustainable. Students typically throw their hands up in agreement.

Then comes the twist: BAT and Glencore – rejected by nearly everyone – receive top ESG scores. Meanwhile, SunPower and Evoqua, widely seen as sustainability leaders, rank among the worst.

Clearly, there is a mismatch between students' perceptions of how sustainable these companies are and how ESG rating agencies evaluate companies!

Admittedly, these examples are overly simplistic, and there are many positive and negative things to say about each of the four companies beyond the snapshots provided. Still, the lecture-hall votes give anecdotal evidence suggesting that ESG ratings may not measure how good or bad a company is for sustainable development.

Clearly, something does not add up. But beyond the smell test, how can we rigorously assess whether ESG ratings capture real sustainability or miss the point entirely?

4.3.4 Putting ESG Ratings to the Test

My former colleague Joop Huij provides equity index investing solutions to clients around the world. When we first met in 2020, he faced a challenge. His client asked for a sustainable index based on ESG ratings. Yet, they were dismayed by some of the companies that made it into the solution that the team developed.

Partly due to their solid ESG rating performance, the index included tobacco companies, mining firms, and fast-food franchises. The client had the same reaction as the students above: while ESG ratings were the market's dominant sustainability metric, they did not seem to really measure the impact that companies have on people or the planet. The client asked why these companies got such good ESG ratings and whether these companies were simply outliers.

Next to being a Managing Director at Robeco, Joop is also an Associate Professor at Erasmus University. His sense for business opportunity combined with his natural curiosity and academic rigour led him to ask me how we could test whether sustainability ratings – like ESG ratings – make sense from a sustainability point of view. As a real quant, he did not think my anecdotal voting among students – or our clients – would suffice. Anecdotes are nice, but we needed evidence.

To move beyond anecdotal insights, we needed a way to test whether ESG ratings actually reflect sustainability performance. But this presented a challenge.

Testing whether sustainability ratings reflect companies' true impact on society and the environment is difficult. The main reason is that a company's actual level of sustainability performance is nearly impossible to quantify. Companies generate numerous, diverse, and potentially conflicting impacts on societies and ecosystems. Evaluating all of these requires interpreting how desirable each of these effects is, involving subjective decision-making.

You may find a lithium miner perfectly sustainable because it delivers key inputs for batteries – thus supporting the energy transition. But your friend may have concerns related to the environmental damage that is caused by the mining process. Both views have a point, and neither is obviously correct.

As a result, no objective list exists that ranks companies by sustainability. Instead, there are multiple viable interpretations of corporate sustainability, often diverging from one another. Without such an objective list, it becomes difficult to put sustainability ratings to the test.

Let me give you an analogy. Suppose you want to test how good corporate earnings estimates are. With such estimates, investments analysts predict a company's future financial performance. Here, an analyst's forecast (ex-ante) can be compared with actual realized earnings after some time has passed (ex-post). This comparison helps reveal how accurate the forecast was.

With sustainability ratings, however, such a test is not possible – we have the forecast (ex-ante rating), but no definitive outcome (ex-post performance) to compare it against.

I explained the challenge to Joop and suggested a path forward: we could test whether ESG ratings align with how investors and regulators judge companies' positive and negative sustainability impacts. If ratings converge with these views, they capture sustainability performance; if not, they fall short. This offers an empirical test, going beyond anecdotes.

We started with negative impacts. Many investors publish exclusion lists naming companies they refuse to finance – those tied to human rights violations, thermal coal, tobacco, and more. I collected lists from 28 global investors to get the investor perspective on a negative impact. Then I added data from the EU Taxonomy, which identifies activities causing significant environmental harm, to portray the regulator's perspective. Together, these gave us a robust sample of companies deemed unsustainable.

I then gathered ESG ratings for these firms and the rest of the investment universe – over 6,600 companies – using four providers to capture their divergence. Regression tests showed the results: ESG ratings do not capture negative impacts. They barely distinguished harmful companies from the broader universe.

Next, we evaluated the positive impact. Investors have long offered thematic funds focused on solutions such as renewable energy, healthcare, and water. To approximate the investor's view of positive impact, I collected the companies held in these thematic funds and treated them as the benchmark set of 'impact' firms. To capture which company regulators consider sustainability leaders, I again used the EU Taxonomy. This time, I identified firms with a substantial share of Taxonomy-aligned green revenues. These companies contribute solutions for climate mitigation and adaptation according to the EU's criteria and therefore served as our sample of positive-impact firms from a regulatory perspective.

Using these lists, I ran the same tests. The outcome was similar: ESG ratings showed limited ability to separate companies with positive impact from the rest, despite some variation across providers.

These results, which were published in an academic journal [96], are important for sustainable investors. They suggest that if you base a sustainable investing solution solely on ESG ratings, you are unlikely to avoid investing in harmful companies. You are also not likely to effectively finance those making positive environmental or social contributions.

What are the implications of these results for sustainable investors? Suppose you are an investor. Your job is to create an investment portfolio that mimics the All Country World Index created by MSCI. This is an index that contains around 3,000 large and mid-sized companies from more than 20 developed and more than 20 emerging markets. It really is a representation of the public companies operating in the world economy.

However, your client does not simply want to invest in all of these companies. They want you to avoid investing in companies that damage the environment or harm societies, while allocating more funds to those that are good for the world. To this end, your strategy is not to invest in the worst-rated ESG companies, using the scoring provided by Refinitiv. You obtained scores for all of the companies and decided to cut the bottom 25% of the firms from your investment portfolio – these are excluded. In other words, you invest in the 75% of the companies with the best ESG rating.

Now how sustainable will your portfolio be? The answer: disappointingly unsustainable.

You have completely failed to avoid companies with a negative impact. Your portfolio includes 73% of the 285 companies in the index that asset owners have put on their exclusion lists for being involved in human rights violations, tobacco or thermal coal production, or other types of sinful behaviour. You now also own a striking 16 of the 17 companies that the European Commission views as causing severe environmental damage. In other words, there is plenty of negative impact on your portfolio.

In terms of investing in solutions, you also do not fare well. You miss 28% of the 458 companies that are included in sustainable thematic portfolios and are thus seen by the market as having positive impacts. Further missing are the 18 out of the 59 companies that have two-thirds or more of their revenues derived from products that the European Commission perceives to be climate solutions. As a result, many companies making real contributions to sustainability are excluded from your portfolio. These patterns are not unique to Refinitiv's rating. Other ESG ratings, like those from S&P, MSCI, and Sustainalytics, also have a high likelihood of missing companies with positive impact and investing in those with harmful effects.

In short, creating sustainable investing portfolios based on ESG ratings not only misses the (sustainability) point, but also brings a high risk of being accused of greenwashing.

4.4 How the Filtering Hedgehog Misleads on Real-World Impact

When I present these findings to experts in sustainable investing, some respond with a shrug. They are not surprised that ESG ratings fail to capture companies' environmental and social impacts. And they are right: most ESG ratings are not designed to measure whether companies contribute to a better world. Instead, they assess whether sustainability issues pose risks to a company's profitability.

In this way, ESG ratings evaluate a company's exposure to, and management of, financially material ESG risks. Most do so by comparing firms within the same sector. So yes, a tobacco company or coal miner can receive a top ESG score – if it manages its risks better than its peers. When such firms are labelled ESG leaders, it is not a flaw in the system. It is the system working exactly as intended: measuring financial risk, not real-world impact.

This is the essence of the filtering hedgehog's approach. ESG ratings may appear 'foxy' in their breadth – covering a wide range of sustainability topics – but they are ultimately driven by one big idea: financial materiality. All environmental, social, and governance factors are filtered through a single lens: how they affect financial performance. Sustainable development, the very goal of sustainable investing, is left out of the equation.

So, is it fair to criticize ESG ratings for missing impact? I believe it is essential. ESG ratings have become the dominant tool for constructing sustainable portfolios. They shape markets, directing capital toward companies – and funds – labelled as leaders. Yet, many of these funds fall short of their sustainability promises, often treating ESG as synonymous with sustainability. It is time to clear up the confusion.[4] ESG ratings can help investors manage risk, but they will not steer capital away from harmful companies or toward those solving sustainability challenges. This makes them insufficient for developing sustainable investing solutions.

[4] This confusion is not accidental. Rating agencies have helped fuel it. Refinitiv claims its ESG ratings help investors 'make sound, sustainable investment decisions'. S&P says its ratings can be used for 'sustainability-linked financing'. MSCI argues its ESG rating supports 'long-term value creation'. The conflation of ESG and sustainability remains widespread in investment practice, media coverage, and academic research.

4.5 The Accountant Hedgehog

In 2013, Peter Bakker – president of the World Business Council for Sustainable Development – famously proclaimed that '*accountants will save the world*' [97]. His idea was simple: businesses will only take sustainability seriously once it is expressed in financial numbers.

This is the realm of the accountant hedgehogs. It seeks to monetize sustainability impacts – turning tons of CO_2, cubic metres of water, or even human rights violations into monetary values. This allows externalities to appear in a profit-and-loss statement, enabling budgets, trade-offs, and performance management. Without this translation, sustainability risks remaining disconnected from the practical reality of how companies and investors act. These are real – and necessary – advantages that can drive positive change.

The Impact-Weighted Accounts Framework (IWAF) is a clear expression of this approach. Its ambition is to '*measure social, human, and environmental impacts and translate them into the language of currency*', creating a harmonized impact language that can sit at the core of strategic decision-making [98]. Supported by institutions such as Harvard Business School and Singapore Management University, IWAF maps impacts across different forms of capital – financial, human, natural, social – and specifies footprint indicators alongside monetization factors that allow translation into financial values.[5]

A case study of Inditex – the Spanish parent company of Zara and the world's largest fast-fashion group – illustrates both the power and the limits of this approach. With over 7,000 stores in nearly 100 countries and annual revenues of nearly €28 billion in 2020, Inditex is one of the most financially successful companies in modern retail. Its business model is built around high-frequency collections produced through an outsourced supply chain in low-cost countries. This model has proven exceptionally profitable but also generates substantial negative impacts well beyond the company's legal boundaries. Supply chain workers face poor wages and poor conditions. Greenhouse gas emissions run across the entire value chain, with 'Scope 3 emissions' – those generated in the supply chain – accounting for roughly 98% of the total. And unsold clothing is frequently burned.

[5] Interestingly, although IWAF enjoys strong support among world-class business schools, it does not mention being supported by sustainability scientists. It is likely that business school professors are more naturally inclined to take an 'accounting approach' to sustainability assessments, compared to those in the natural or social sciences.

Sustainable finance researchers Dirk Schoenmaker and Willem Schramade applied the IWAF methodology to calculate not only Inditex's financial value but also its social and environmental value, converting these impacts into monetary terms [99].

The result, at first glance, looks reassuring: Inditex's overall integrated value – financial, social, and environmental combined – turns out to be positive. But the headline number obscures what lies beneath it. The negative environmental and social impacts, once monetized, amount to -€320 billion. The climate contribution alone comes to -€101 billion, driven primarily by the Scope 3 emissions that the company's own sustainability targets largely sidestep. By putting a financial value on environmental and social impacts, the case makes clear that Inditex is far less positive than a purely financial assessment would suggest – even if its overall integrated value still comes out positive. But that result depends entirely on the prices assigned to each impact. As the authors note, these numbers are very rough estimates, meant to illustrate the scale of externalities that most companies and investors currently ignore.

This raises an uncomfortable question: who determines the right price for environmental and social externalities?

Examples from the IWAF methodology are striking. A non-fatal workplace injury is valued at $4,360 if insured, $4,550 if not. A fatal incident? $3.7 million. Emitting a kilogram of CO_2 costs $0.236. Polluting air with PM2.5 costs $78.50 per kilogram – $10 cheaper than causing ozone pollution. A child in hazardous work is priced at $36,600. Forcing people to work ranges from $18,000 to $43,000 depending on conditions. Even life satisfaction is given a price: $2,647 per point on a 0–100 scale. The list goes on.

These numbers aim to make the invisible visible. But they raise hard questions. Who decides the 'right price' for clean water, cultural heritage, biodiversity, or a human life? IWAF explains its methodology, yet subjectivity is unavoidable. As Robert Shrimsley quipped in the Financial Times [100] after a study priced UK parks at £34.2 billion a year, or £30.24 per person: is not his enjoyment worth £32.56? And is not the beautiful wisteria, currently valued at 8.3p a day, worth 22p at least? His satire underscores the tension: can life satisfaction, nature, or dignity be reduced to a spreadsheet?

This is more than a technical debate. It cuts to the heart of the hedgehog–fox divide: whether the many dimensions of sustainability can ever be reduced to a single price.

4.6 How the Accountant Hedgehog Creates an Illusion of Comparability

The accountant hedgehog's ambition rests on a bold assumption: that all impacts can be converted into a single currency and compared on equal terms. This is where the fox raises its sharpest objection.

As we know, sustainability is not one thing; it is many things. A company's footprint spans social values like employment, healthcare, and human rights, and ecological values like biodiversity, water quality, and carbon emissions. These values are not naturally commensurable. They cannot be neatly transformed into quantities that can be traded off against one another.

This becomes painfully clear when values conflict. A security company may enhance public safety while eroding individual privacy. A lithium miner may help mitigate climate change while destroying fragile ecosystems. Both sides matter, yet maximizing one diminishes the other. Converting these impacts into money creates the illusion that they can all be optimized at once, obscuring the real trade-offs that sustainability demands.

The problem runs deeper. Even when values do not directly conflict, monetization can flatten their meaning. Imagine a company that produces nutritious food – a positive contribution – while relying on child labour – a profound negative. Once these impacts are expressed in dollars, the company's overall score can improve simply by selling more healthy food products, even if it never addresses child labour. The plurality of values like clean energy and decent work disappears in the hedgehog's monist worldview.

And then there is the arithmetic of scale. The Inditex case illustrates this vividly. Its overall integrated value comes out positive – not because its environmental and social harms are small, but because large positive values such as consumer surplus, employment, and tax payments arithmetically offset them. By reducing everything to money, the system privileges firms that are financially succesful. For a company earning billions, a $3.7 million 'cost' for a fatal accident is a rounding error, and high tax payments and lavish salaries can easily outweigh environmental destruction or human rights violations on the balance sheet. The logic of monetization does not just obscure trade-offs, but risks legitimizing harm.

The researchers themselves warn against this, arguing that their results show the importance of not netting the values, since both the positive and negative social and environmental figures are far larger than the company's financial value [99]. The large negatives need to be addressed – they cannot simply be cancelled out. This is not to say the exercise is futile. Even rough, contestable prices can reveal the sheer scale of externalities that financial accounts ignore entirely.

Ultimately, this is the fox's concern: sustainability is not a single number. It is a tapestry of values that resists being collapsed into one metric. When we force commensurability, we lose sight of what makes these values matter in the first place.

4.7 Concluding Remarks: Towards Foxy Sustainability Screening

The hedgehog has long dominated sustainable investing. Its tactics – whether specializing in a single issue like emissions, filtering multiple issues through a singular lens like ESG ratings do with financial materiality, or accounting for impacts in monetary values – offer simplicity, structure, and a sense of control. But these tactics are insufficient. They often ignore the messy, interconnected reality of sustainability and risk misaligning investment strategies with the very goals they claim to support.

The singular screens used by the specialist hedgehog, no matter how well-intentioned, can lead to blinkered portfolios that wilfully ignore impacts on other relevant sustainability topics. In turn, despite the breadth of topics considered by the filtered hedgehog, its ESG ratings do not measure companies' impacts. And when the accountant hedgehog turns sustainability impacts into currency, it obscures rather than clarifies. In all cases, the hedgehogs' approaches miss the point: sustainable investing is not just about tackling a single sustainability issue, improving financial returns, or translating impacts into money – it is about contributing to sustainable development in all its diverse facets.

This misalignment is not merely academic. It affects how trillions of dollars are allocated, which companies are financed, and which solutions are scaled. Investors may want to support sustainability, but if the tools they rely on are flawed, their portfolios may end up financing harm or overlooking impact. This is manifest in real portfolios. Climate funds investing in companies that undermine wellbeing, or health funds financing firms that exacerbate global warming, are not uncommon. Such misalignments not only contradict sustainability science – they also expose investors to reputational risk and 'greenwashing'.

To move forward, we need to rethink how sustainability screening is done. That means embracing complexity, acknowledging trade-offs, and seeking synergies. It means moving beyond monistic metrics and toward pluralistic assessments. In short, it means making sustainable investing foxy.

In the next chapter, we will explore what foxy sustainability screening might look like and how investors can build portfolios that reflect the plural, interconnected nature of sustainable development.

5

Making Sustainable Investing Foxy

How to Create Effective Sustainable Investment Strategies?

At some point, two sustainability screening methods crossed my desk. Both were developed by the same organization. One assessed how companies address climate change; the other evaluated their engagement with biodiversity. The approaches were well presented in shiny slide decks, and a case was made that they were truly innovative. Against a background of existential threats coming from unmitigated climate change and collapsing biodiversity, this seemed worth exploring.

But when I examined the results, the disconnect between the climate and biodiversity screens was striking. A major oil company was flagged as a poor performer on climate – sensible enough, given its role in expanding oil production. But the same company was rated a leader on biodiversity.

This is not just counterintuitive. It is scientifically indefensible. Climate change is one of the leading drivers of biodiversity loss worldwide. A company causing the former logically also generates adverse impacts on the latter. The hedgehog's single lens had produced a contradiction that basic sustainability science would never permit.

After years of encountering such approaches – each with its own blind spots, each confidently presenting scores that obscured more than they revealed – I have come to believe that sustainable investing needs a fundamentally different way of thinking. It does not need a better hedgehog. Instead, it needs a fox.

Unlike the hedgehog, the fox thrives on adaptability. It does not cling to one big idea but survives by using a range of tactics and strategies. Nature offers striking examples: the arctic fox, the best-insulated mammal on earth; the fennec fox, built to endure Saharan heat; and the common red fox, a rare

J. A. van Zanten, *Big Money, Better World?*,
https://doi.org/10.1007/978-3-032-32442-9_5

carnivore that has learned to thrive in city centres. The fox's strength lies in flexibility, reading its environment and adjusting accordingly.

In sustainable investing, the fox is in many ways the opposite of the hedgehog. The hedgehog clings to a single lens – often financial materiality or one dominant sustainability issue – and builds its entire approach around it. The fox, by contrast, embraces complexity. It does not seek a supreme metric or one overarching goal. Instead, it values the plural and interconnected nature of sustainable development. Foxes make decisions by drawing on interdisciplinary insights, adapting to new evidence, and recognizing that sustainability challenges are multifaceted and constantly evolving.

For investors, this foxy mindset can feel daunting. The landscape of sustainable investing is crowded with metrics, ratings, scores, frameworks, and datapoints. It is easy to feel lost, uncertain, or even overwhelmed. How can investors tell whether a sustainability screening tool reflects the fox's approach, and whether it is useful for building effective investment strategies?

What is needed is not just more screening methods. What is needed is a way to make sense of them: to evaluate their adequacy and determine whether they help investors support sustainable development.

This chapter aims to do just that. It sets out three principles that effective sustainability screens should meet, then presents a tool that comes close to meeting them. It then asks why such approaches have not yet become the standard – and explores the frontier of new research pushing them further. Finally, it shows how these principles can be applied in practice and what investors must keep in mind when building sustainable portfolios.

5.1 Principles for More Effective Sustainability Screens

As argued earlier, sustainability screening is a core element of sustainable investing. An effective screening method helps investors understand how sustainable companies are. Based on that understanding, investors can allocate capital toward leaders and away from laggards – and use their voice to push for positive change through active ownership.

So, what general principles should effective sustainability screens adhere to? Here are the three I believe are most important for developing a foxy approach.

5.1.1 Evaluate a Company's Plural Role in Sustainable Development

The primary objective of an effective sustainability screen is to evaluate a company's role in sustainable development. This requires assessing how companies perform across a wide range of environmental, social, and economic sustainability issues – simultaneously. In other words, the screen must have a plural scope that identifies which sustainability goals a company helps advance and which it may hinder.

This pluralism is at the heart of the fox's approach. Rather than reducing sustainability to a single metric or overarching goal, the fox evaluates companies based on their contributions – positive and negative – to many sustainability issues. As discussed in Chapter 2, the most comprehensive framework for defining these issues is the Sustainable Development Goals (SDGs), adopted by the United Nations. These 17 goals and 169 targets offer a structured, albeit imperfect, way to scope sustainability performance.[1]

By screening companies on their impact across multiple sustainability objectives, we avoid the hedgehog's traps of looking only at a single issue and of imposing a singular yardstick – like ESG ratings do with 'financial materiality'. The foxy approach resists this reduction of looking at a grand overarching view. It measures impacts on specific sustainability goals and leaves it at that.

5.1.2 Determine a Company's Overall Contribution

Evaluating a company across multiple sustainability issues is essential, but it is not enough. An effective sustainability screen must go further and deliver a clear judgement on the company's overall contribution to sustainable development. Without this, investors face confusion: how should they interpret a company that performs well on some issues but poorly on others?

To provide clarity, the fox must make trade-offs. Most companies have both positive and negative impacts across different sustainability dimensions. Weighing these impacts and determining an aggregate assessment is complex but necessary.

[1] The SDGs have drawn various criticisms. Some argue that there are too many goals and targets without a framework for achieving them, making it hard to guide policies. Others focus on the content of the goals, positing that the economically oriented goals are inherently at odds with the environmental objectives – thus presenting a contradictory agenda. And another concern is the fact that the SDGs lack legally binding enforcement, which made it easier for them to become universally adopted, but now raises concerns over greenwashing.

A critical principle in making these trade-offs is the primacy of avoiding harm. Advancing sustainable development starts with preventing substantial negative impacts before rewarding positive ones.[2] In practice, this means that companies causing severe harm should receive a low overall rating, regardless of any positive contributions they make elsewhere. Conversely, companies with limited negative impacts can score more positively, with higher ratings reflecting the extent of their beneficial contributions.

This logic ensures that positive impacts in one area cannot compensate for significant harm in another. A company selling nutritious agricultural products harvested through child labour is, overall, unsustainable. This gives the fox a strong normative foundation and avoids sliding into relativism – the idea that all values are subjective and interchangeable. Instead, the fox adopts an objectivist approach anchored in evidence-based principles.[3] This sets the stage for the next criterion.

5.1.3 Build on Sustainability Science

To judge how desirable a company's impacts are – on individual themes and overall contributions – the fox must anchor its approach in sustainability science. Scientific research offers evidence on how corporate operations and products affect sustainability objectives, enabling informed distinctions between impacts.

In one of the papers I wrote for my PhD, I reviewed nearly 900 articles to understand how specific corporate activities impact different sustainability topics [101]. This now helps me to screen companies on their impacts and advise investors on which to chase and which to avoid.

Consider the following examples. In the energy sector, scientific evidence shows that while all fossil fuels drive climate change, thermal coal has by far the greatest negative impact, followed by oil, and then natural gas. Similarly, in the food sector, beef production generates substantially higher

[2] This is known as the precautionary principle – which has gained wide traction in many disciplines, including environmental policy.

[3] For those with an interest in philosophy, it is worth underscoring that foxes are not relativists. Relativists deny the existence of 'real' moral values, believing people can simply make up whatever values they like. Pluralists (the foxes), by contrast, affirm that real moral values exist – but unlike value monists (the hedgehogs), they hold that these values are plural rather than reducible to one supreme principle. In the context of sustainable investing, this means the fox acknowledges that multiple sustainability goals matter and must be pursued. Even so, the fox cannot act arbitrarily. It must reach an overall judgement of how good or bad a company is, based on evidence and reason. Sometimes that judgement will be wrong – underscoring the objective component of value pluralism.

environmental impacts – emissions, land use, and water consumption – than poultry, while plant-based products generally have the lowest adverse effects. And whereas solar energy and bioelectricity are both solutions for mitigating climate change, the former has a far greater potential contribution than the latter [102]. A credible scoring system must reflect these differences rather than treating 'fossil fuels', 'food products', or 'renewable energy' as homogeneous categories.

Ideally, the impacts of corporate products and operations should be evaluated in relation to the planetary boundaries that define safe operating limits for humanity [103–105]. For example, a firm extracting water in arid regions has a far more detrimental effect on local ecosystems than one using water in areas with an abundant supply. Sustainability science can help determine how companies relate to these boundaries, even if assessing such relationships remains complex and, in many cases, methodologically challenging.

But sustainability science is not merely a catalogue of findings about isolated sustainability issues. Its deeper contribution lies in revealing how those issues are connected. Disruptions in one part of a social-ecological system – the interconnected web of human activity and the natural environment – propagate through others, generating ripple effects that no single-issue screen can detect.

This is what makes the opening example of this chapter so instructive. A company dedicated to pumping oil directly drives climate change. Climate change, in turn, causes wide-ranging harm to biodiversity, both on land and below water. Further still, these ecological disruptions feed back into social systems: heatwaves, droughts, floods, and land degradation affect people's lives and livelihoods. Climate change, biodiversity loss, and human wellbeing are therefore not parallel concerns – they are entwined components of the same system. Seen this way, the oil company's contradictory scores were not merely counterintuitive. They were a scientific impossibility.

Addressing this blind spot requires systems thinking – and sustainability science calls for exactly that [106–108]. Rather than examining sustainability challenges in isolation, a system-based approach maps how they interact: where progress on one goal creates co-benefits for another, and where contributions to one goal risk undermining another. When a company causes harm through one dimension of its operations, that harm rarely stays contained. It propagates through ecosystems, supply chains, and communities in ways that only a systemic view can trace. This also explains why the primacy of avoiding harm is not merely a normative preference, but a scientific one: significant negative impacts tend to be systemic, and no amount of positive performance elsewhere can undo their cascading effects. The same logic also surfaces opportunities. A firm providing plant-based protein for replacing meat consumption may well help reduce deforestation, which indirectly advances climate, biodiversity, and water security goals – relevant co-benefits that a siloed screen would never capture.

This principle of being built on sustainability science – and through systems thinking – also means screening cannot be static. It must evolve as scientific knowledge advances, continuously integrating the latest insights on how companies and their products affect the social and ecological systems on which we all depend.

5.2 From Principles to Practice

The principles introduced above ensure that a foxy sustainability screen does three things: it evaluates companies across multiple sustainability goals, delivers an overall judgement that prioritizes avoiding harm, and grounds its approach in the best available science. These principles ensure that screening methods reflect the complexity of sustainable development, provide clarity for investors, and remain credible as knowledge evolves. In short, they help move sustainable investing beyond hedgehog simplicity toward a more adaptive, evidence-based approach.

So, what would following these principles look like in practice? One example comes from SDG scores, which have gained prominence in the sustainable investing industry in recent years.

5.2.1 SDG Scores

SDG scores aim to measure how companies impact the Sustainable Development Goals. One of the most prominent examples is the Robeco SDG Score, based on a framework I helped design and scale.

The idea took shape in 2017 through a confluence of frustrations and ambitions. Taeke Wiersma, then Head of Credits Research at Robeco, had grown increasingly dissatisfied with ESG ratings. He felt they failed to capture companies' real-world impacts and were misaligned with clients' normative views on sustainability. Robeco's sustainable investing research team, led by Jake Messina, shared the same concern.

Around the same time, Victor Verberk – Robeco's Chief Investment Officer – attended an investment conference where SDG logos featured prominently across presentations and booths. It struck him: this was exactly the kind of language his clients understood and cared about. He wanted to offer SDG-aligned strategies, but quickly ran into a problem: no metrics existed to assess corporate alignment in a rigorous way.

The three converged on the same conclusion from different angles. The SDGs offered a compelling blueprint for a new kind of screening method, yet turning that blueprint into something a portfolio manager could actually use was another matter entirely. A structured, rules-based methodology was essential. Without it, assessments would come down to personal judgement, producing inconsistent opinions, internal debates, and results that could not be replicated. Victor backed the initiative on one condition: the methodology had to be rules-based.

Over time, the team, supported by a small army of the firm's credit and sustainability analysts, developed a framework to assess companies' impacts on the SDGs in three steps: evaluating the impact of a company's products and services, its operations, and any controversies it may be involved in.

Companies are scored across all 17 SDGs, from –3 (highly negative) to + 3 (highly positive). These are then aggregated into a total SDG score that signals overall alignment.

The first step evaluates the impact of a company's products and services. Renewable energy, clean water, and healthcare technologies score positively; gambling equipment, thermal coal, and cluster munitions score negatively. To ensure consistency, the framework uses revenue thresholds to assign scores. For example, beef, lamb, and mutton – linked to water scarcity, land degradation, and greenhouse gas emissions – receive negative scores based on revenue share: –1 for 20%, –2 for 40%, and –3 for 60%. This logic applies across sectors.

The second step assesses operations, covering universal dimensions like governance and gender equality, as well as sector-specific issues such as sustainable palm oil sourcing or aquaculture practices. Again, clear rules determine what level of performance earns a given score.

The final step checks for controversies. Companies are penalized if they are involved in serious environmental or social scandals, fail to remediate harm, or lack preventive measures.

Once all three steps are complete, companies receive scores on each SDG. The framework focuses on substantial impacts, not peripheral efforts like volunteering or donations. As a result, most companies score positively or negatively on two to four SDGs and neutrally on the rest.

Each company then receives a total SDG score, applying a 'min–max' rule: if a company has a negative score on any SDG, that (minimum) score becomes its total score. Only companies with neutral or positive scores across all SDGs earn the highest rating.

To illustrate: a solar panel manufacturer may score well for its contribution to the energy transition. If its operations and conduct are otherwise sound, it receives a positive total score. But if the company uses forced labour in its supply chain, it earns a negative score on SDGs related to decent work, and that negative score becomes its total score. This foxy logic reflects a core principle: positive impact on one goal cannot offset harm on another.

So, what does this assessment reveal? Robeco's analysis of nearly 3,000 companies worldwide shows that 27% of companies score negatively – indicating they contribute to harmful effects that undermine global sustainability progress. Just over half (53%) receive a positive score, with around 7% earning the highest possible score. The remaining 20% are rated neutral, meaning they do not significantly impact the global goals [109].

Beyond total scores, the framework highlights which SDGs receive the most positive and negative support. The strongest positive contributions are linked to SDG 8 (Decent Work and Economic Growth) and SDG 9 (Industry, Innovation, and Infrastructure), with roughly one-third of companies making meaningful contributions. Other frequently supported goals include SDG 11 (Sustainable Cities and Communities) and SDG 3 (Good Health and Well-being). Unsurprisingly, the most negative impacts cluster around SDG 13 (Climate Action), with 12% of companies – such as fossil fuel producers, car manufacturers, airlines, and cruise operators – causing harm. SDG 15 (Life on Land) also sees significant negative exposure, particularly from unsustainable palm oil sourcing and fertilizer production. SDG 16 (Peace, Justice, and Strong Institutions) is harmed by about 7% of companies, often due to involvement in controversial weapons, human rights abuses, bribery, or corruption.

The SDG score reflects the three criteria outlined earlier: it has a plural scope, calculates an overall score that avoids harm, and is informed by sustainability science. As a result, it offers a more nuanced approach to sustainable investing. Unlike the hedgehog's singular focus, the fox's methodology recognizes that sustainability is multidimensional, filled with trade-offs and synergies. By evaluating companies across multiple goals and applying rules that prevent harm in one area from being offset by progress in another, these frameworks align more closely with the complexity of real-world sustainability. They help investors identify true leaders and avoid false positives; companies that look good on one metric but undermine progress elsewhere.

This SDG score gained wide traction. When I left the firm in 2025, tens of billions of euros in assets were influenced by it. Next to investors, the score found traction in academic research. Various publications featured it as a measure of sustainability performance, leading to new insights on corporate sustainability and sustainable finance.

Other organizations have developed similar SDG scoring systems, including MSCI, ISS, and a coalition of asset owners including PGGM, APG, AustralianSuper, and BCI. These ratings are gaining real traction across the industry.[4] This shows that many in the investment industry are exploring how firms impact global goals like the SDGs.

At this stage, you might wonder whether SDG scores suffer from the same flaws as ESG ratings. The short answer: they perform much better. For one, SDG scores are not influenced by company size, location, or disclosure practices, thus avoiding the biases that plague ESG ratings [110]. That said, scores from different providers do diverge [111]. But this divergence may not be a flaw. In fact, my research showed that SDG scores from different providers appear effective at identifying companies with positive or negative impact [96]. This matters: even when raters disagree on the details, they converge on which companies are true leaders or laggards in contributing to sustainable development.

5.2.2 The Fox in Action

The principles of foxy investing – plural scope, overall contribution, and sustainability science – are easy to endorse in the abstract. But do they actually produce different results in practice? Two examples show how the fox's approach leads to materially better judgements than any hedgehog alternative.

The first is BP. Those two screens from the opening of this chapter – the ones that rated a major oil company as a climate laggard and biodiversity leader – were applied to BP. That result, as I argued there, is scientifically indefensible. But it illustrates something important: the hedgehog's failure is not accidental but structural. When you assess climate and biodiversity through separate, single-issue lenses, you lose the connections between them. And in sustainability, those connections are everything.

[4] Although these are all examples of sustainability screening methods that evaluate corporate impacts across a plurality of topics – the SDGs – not all meet the three principles set out in this chapter. Some do not generate an overall judgement on a company's sustainability performance. Others allow negative impacts to be offset by positive ones, undermining the primacy of avoiding harm. And still others pay limited attention to adverse impacts altogether, rewarding positive contributions without adequately penalizing the damage a company causes.

The fox has no such problem. Assessed simultaneously across climate, biodiversity, labour rights, governance, and more, BP's picture is consistent: a company whose core business generates substantial negative impacts across several deeply interconnected goals. BP's worst scores define its overall rating – no positive grade elsewhere can offset that.

The practical consequence is real. A biodiversity-focused hedgehog screen might see oil companies like BP as leaders and include it in sustainable investing strategies. A foxy screen would recognize that its adverse climate impacts inevitably degrade biodiversity too, making it a poor candidate for any sustainable portfolio.

A second example is Siemens, which presents the mirror image of the BP problem. The German industrial giant has a substantial carbon footprint. This is primarily a consequence of its manufacturing operations, including the production of wind turbines, which are energy-intensive to produce. A specialist hedgehog focused on greenhouse gas emissions would penalize Siemens, potentially excluding it from a sustainable portfolio.

But this would be a serious mistake. Siemens is among the most innovative companies in the sustainability transition. Its portfolio includes wind energy infrastructure, smart grid technology, energy-efficient building systems, and industrial automation that reduces emissions across entire supply chains. The products it manufactures – even the carbon-intensive ones – enable other companies and economies to decarbonize at scale. In other words, Siemens provides solutions that are necessary for tackling climate change, even though these come with a substantial carbon footprint.

A foxy screen, grounded in sustainability science and assessing companies across multiple goals, captures this complexity. Unless issues with substantial negative impact emerge elsewhere, Siemens scores positively for its contributions to the energy transition, thereby signalling a company that advances sustainable development and deserves investor support.

A hedgehog investor focused on reducing portfolio emissions might reach the opposite conclusion, excluding Siemens on the basis of its manufacturing footprint alone. The irony is sharp: by doing so, they finance fewer solutions to the very problem they set out to address. The fox avoids this trap entirely.

Together, BP and Siemens illustrate what the fox's approach achieves that no hedgehog can: it produces judgements that are coherent, evidence-based, and resistant to the distortions that single-lens thinking inevitably introduces. The hedgehog creates two types of error. It excludes companies that deserve inclusion – like Siemens, penalized for the manufacturing cost of solutions it provides. And it includes companies that deserve exclusion – like BP, cleared on biodiversity despite being a driver of biodiversity loss.

These are not edge cases. They are symptoms of a systematic flaw. The difference between hedgehog and fox thinking is ultimately the difference between sustainable investing that really contributes to a better world and sustainable investing that merely looks like it does.

SDG scores, as we have seen, go a long way toward making that distinction. But they are not the final word. Before we go into the frontier of sustainability screening methods, let us examine why foxy approaches to sustainable investing are not widespread.

5.3 What Is Holding the Fox Back?

Why do hedgehog approaches to sustainable investing still reign supreme? A simple answer is that the hedgehog's appeal lies in its simplicity. Hedgehogs focus on a single issue, apply one dominant lens to many issues, or compress diverse impacts into a single value. Foxes, by contrast, embrace plurality. Their approach requires understanding how companies perform across multiple, interlinked sustainability challenges – and then forming a coherent overall judgement. That is harder. It demands more data, more expertise, and more judgement. It also requires investment. Unsurprisingly, truly foxy approaches remain relatively rare in practice.

But complexity alone does not explain the fox's slow uptake. Something more fundamental is at play: a principal-agent problem between asset owners and the asset managers who invest on their behalf.

Many asset managers are primarily evaluated on one dimension: their ability to generate alpha. Outperforming benchmarks is the industry's dominant scorecard. Within that logic, sustainability is often perceived as a constraint, as something that narrows the opportunity set and potentially interferes with return generation.

This perception is reflected in how asset managers themselves describe their motivations. A recent academic study asked over 500 fund managers about sustainable investing [112]. Financial returns were the dominant objective for over two-thirds of respondents. Only 27% said they would tolerate companies sacrificing even one basis point (0.01 percentage point) of annual return in exchange for better sustainability performance. Most pointed to regulatory or legal constraints, arguing that such trade-offs would conflict with their fiduciary duty to clients.

This return-first mindset sits uneasily with the preferences of many asset owners. Pension funds, insurers, and other long-term investors increasingly want their capital to do more than generate financial returns alone. Because they invest over decades, supporting sustainable development becomes a legitimate objective alongside risk and return – not a distraction from it.

Survey evidence supports this divergence. Asset owners are, on average, more willing than professional money managers to accept modest financial trade-offs in exchange for stronger alignment with social and environmental goals.

One particularly revealing study examined the preferences of individuals whose money is managed by a Dutch pension fund with nearly EUR 21 billion in assets [113]. In a survey of 1,700 of its beneficiaries – whose responses would directly shape the fund's investment policy[5] – two-thirds wanted their pension fund to increase its contributions to societal and environmental goals like those defined by the SDGs, even if it hurt returns. Only 11% opposed the idea, and 21% were neutral. This gave the fund a strong mandate to expand its sustainable investing strategy.

But here is the tension. As the central message of this chapter makes clear, pursuing social and environmental objectives effectively requires robust, nuanced sustainability screens – not simplistic ones. Yet asset managers are often drawn to exactly those simplistic screens, because they are the cheapest to build and they interfere least with conventional investment processes and performance targets. They still allow managers to tick green boxes, while the resulting portfolios may fall well short of what asset owners actually intend.

So, what can be done to ease this principal-agent problem, and, in doing so, make the fox's strategy more mainstream? Part of what sustains this gap is imperfect information on both sides.

Asset owners rarely have full visibility into how sustainability screens used by their managers actually work. Many are marketed elegantly, presenting polished scores and rankings that look convincing at first glance. But understanding how companies are assessed across sectors, regions, and sustainability issues requires time and expertise. As a result, asset owners may struggle to judge whether a manager's screening approach truly aligns with their sustainability goals. Conversely, asset managers may lack clarity about the precise trade-offs their clients are willing to make.

Transparency is a powerful antidote to this information gap.

[5] This is an important feature of the study. Since the responses that were collected would impact the investments of the people that were surveyed, the authors found a way to make it more likely that truthful answers were obtained and that the study paints an accurate picture of sustainable investing preferences.

For asset managers, this means clearly explaining how sustainability tools work – moving away from opaque, black-box ratings – and showing how portfolio holdings are evaluated in practice. For asset owners, it means assessing managers not just on performance, but on the quality of their sustainability screening. Do their methods reflect the principles set out in this chapter? Do they evaluate a plurality of sustainability issues? Do they define an overall contribution that prioritizes avoiding harm? And are they grounded in sustainability science?

Without such transparency, foxiness remains risky. With it, the fox has a chance to become the new norm in sustainable investing. Let us now see what the frontier of sustainability screening research looks like.

5.4 Next-Generation Sustainability Screens

SDG scores are a significant step forward, and I say that with the bias of someone who helped build them. But there is room for improvement.

Most SDG scores focus on measuring companies' current impacts. Yet a growing consensus holds that a forward-looking perspective is an essential addition. Achieving a sustainable future requires not only rewarding today's leaders but also enabling laggards to improve.

Scholars have warned that strategies that simply reallocate capital from brown to green firms risk reinforcing the status quo [114]. The logic is simple: raising the cost of capital for brown firms makes it harder for them to invest in sustainability improvements. But creating a more sustainable future depends on unsustainable companies transitioning to sustainable business models. That means capital should also flow to brown firms that are credibly working to become green. These companies offer potential for transformative impact.

But there is a clear challenge. How can investors identify which companies are credibly transitioning toward better sustainability performance?

If you allocate capital to brown firms, you run the risk of subsidizing business-as-usual rather than enabling real change. If it turns out these firms are not improving their sustainability performance, you have simply reinforced brown activities. And with greenwashing still pervasive, relying on corporate reporting, targets, or stated ambitions is insufficient. They have too often proven unreliable.

However, currently, few screening methods assess whether companies are improving across multiple social and environmental dimensions in an integrated way.

That gap – between what sustainability screens currently measure and what investors truly need to know – is part of what prompted Victor to establish Osmosis Investment Management NL in 2025, bringing together a team of 15, including Taeke and myself. Building a new asset management firm from scratch is not a trivial undertaking. But we saw a real opening: sustainability data is becoming richer and more granular, and advances in computing power and AI are making it possible to analyze that data in ways that were simply not feasible a few years ago. Combining these two developments – better sustainability intelligence and more powerful research tools – creates a real opportunity to push the frontier of what screening can do. This ambition shapes how we invest: we want to finance companies that are already sustainable today, because they deliver the solutions societies need, and firms that are credibly transitioning, because we want to support companies not just for where they are, but for where they are going.

Let me give you some examples of what we are looking at.

5.4.1 Sustainable Revenue Momentum

One dimension we investigate is what we call sustainable revenue momentum. We screen all companies on their revenue mix, identifying what percentage of their sales derives from each of nearly 2,000 products and services. For a mining firm, this means knowing the share of revenues from copper, nickel, zinc, lithium, and coal. For a healthcare company, it means distinguishing between pharmaceuticals, biotech, and diagnostics.

This granularity matters because products vary widely in their environmental and social impacts. We have evaluated each of these nearly 2,000 products and services on their impact magnitude – the extent to which they generate positive or negative effects across environmental and social dimensions. Knowing a company's revenue split therefore tells you something important about where its impact stands today, as captured by the SDG scores discussed above.

But by tracking revenue mix over time, we can go further. As a company's portfolio of products shifts, so does its sustainability profile. Growing green revenues suggest a firm is transitioning toward more positive impact; rising brown revenues signal the opposite. This gives investors a glimpse of where a company's impact is heading – not just where it stands.

Two real examples illustrate what this looks like in practice.[6]

[6] I am grateful to Marit Piek for her research support. Between January and April 2026, she did an internship with us at Osmosis NL, conducting a thorough assessment of this sustainable revenue momentum factor and developing the insights presented here.

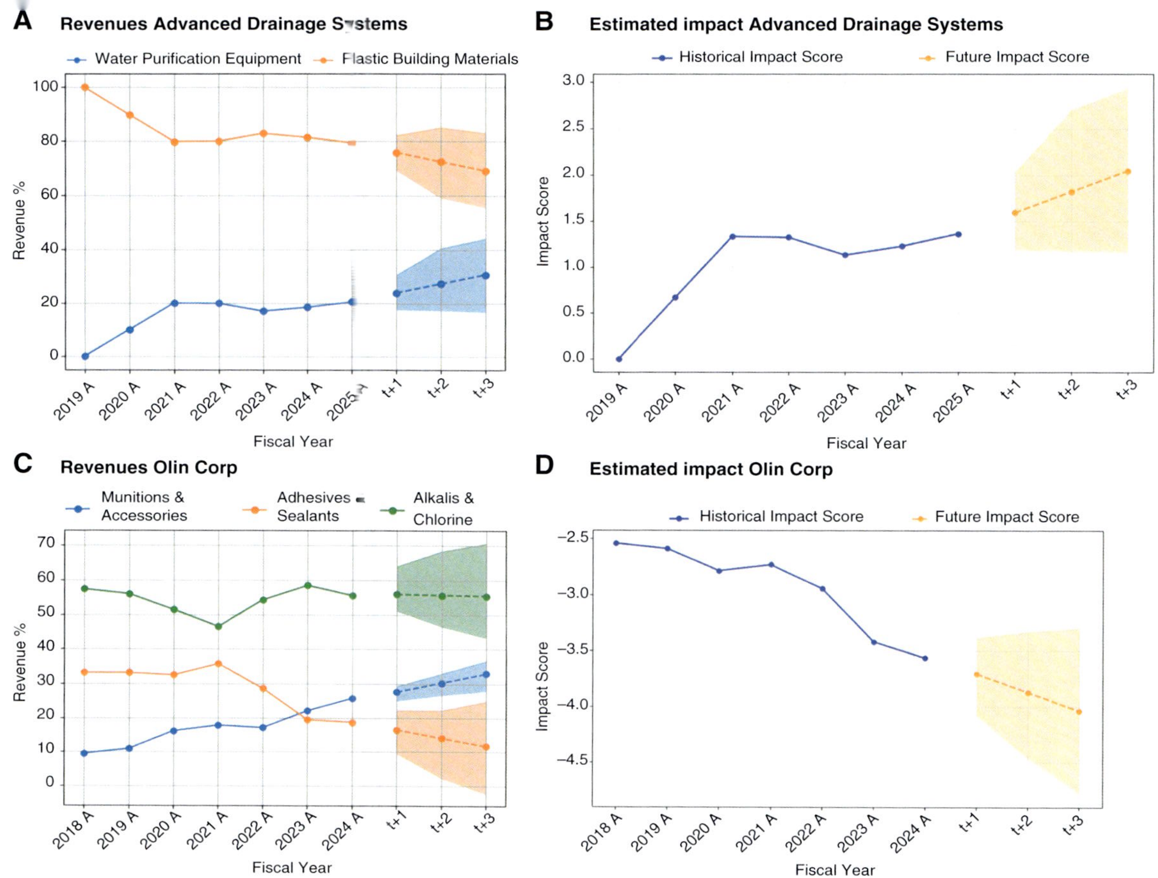

Fig. 5.1 Illustration of sustainable revenue momentum. This figure illustrates a sustainable revenue momentum analysis for two companies. Panels A and B describe, respectively, changes in revenue segmentation and related impact scores for Advanced Drainage Systems. Panels C and D do the same for Olin Corp. Solid lines show actual historical data while dotted lines show predicted data.

The first is Advanced Drainage Systems, a company producing water management solutions. In 2019, all of its revenues came from plastic building materials. Since then, it has steadily expanded into water purification equipment, which accounted for just over 20% of sales by 2025. Panel A in Fig. 5.1 shows this revenue shift over seven years, drawn from the company's annual filings to the U.S. Securities and Exchange Commission, alongside a three-year revenue forecast and confidence interval.[7]

These two product categories carry different impact profiles in our model. Plastic building materials we rate as neutral; water purification equipment as medium positive. As the revenue mix has shifted, so has the company's overall impact score – shown in Panel B. On a scale from –10 (highly negative) to + 10 (highly positive), Advanced Drainage Systems scored around 1.5 in 2025. Based on projected revenue trends, our model expects that to rise to approximately 2.0 within three years.

The second example runs in the opposite direction. Olin Corp, a producer of chemicals and ammunition, reported revenues across three categories with meaningfully different impact profiles: adhesives and sealants, which we rate as neutral; alkalis and chlorine, slightly negative; and munitions, medium negative. As Panel C shows, munitions have grown as a share of revenues over time, displacing the more neutral adhesives business. The consequence, shown in Panel D, is a sustainability profile moving in the wrong direction – an expected impact score that has deteriorated and is projected to continue doing so.[8]

5.4.2 Sustainable Capital Expenditure

Another dimension is *sustainable capital expenditure (CapEx)*. Firms are increasingly reporting what share of their CapEx is contributing to environmental or social goals.

[7] The forecasts were made using *Holt's forecasting method*, also known as double exponential smoothing. This method is well-suited to data exhibiting modest trends. In essence, the model extends the most recently observed trend as a straight line by placing greater weight on more recent observations than older ones. The determination of these weights is achieved in a statistically optimal way through the estimation of smoothing parameters. The method also produces confidence intervals, which quantify uncertainty in the prediction.

[8] An important caveat: Revenues are an indirect proxy for the volume of goods and services a company actually delivers. A firm's revenue mix may shift due to external factors – commodity price cycles or geopolitical conflicts, for instance – while its real-world output in volume, and therefore its actual impact, remains unchanged. Despite this limitation, sustained revenue shifts over time can signal meaningful change in a company's sustainability trajectory.

To illustrate, energy firm Neste reports that 59% of its 2024 CapEx was dedicated to manufacturing biofuels, and therefore being sustainable in line with the EU Taxonomy [115]. TotalEnergies reported sustainable CapEx of 25%, mostly in renewable energy generation but also in battery manufacturing and new energy infrastructure for low-carbon mobility [116]. And Shell reported a figure of 6%, with investments in solar and wind energy, hydrogen, and infrastructure enabling low-carbon transport, among others [117].

By analyzing sustainable CapEx data relative to a firm's broader investment programme, we can identify which companies are most strongly committing capital to a better future – and therefore which are most likely to transition toward greater sustainability.

5.4.3 Sustainable Innovation

A final dimension worth exploring is sustainable innovation. Revenue momentum and capital expenditure both reveal where a company is heading in the near term – the former through its current product mix, the latter through its investment commitments. But neither captures what a company is inventing for the longer term. Patents offer exactly that: a window into the pipeline of products and processes that do not yet generate revenues or attract capital, but may define a company's sustainability profile a decade from now. By analyzing a company's patent portfolio, investors can gain early insight into its capacity to develop sustainability solutions – or, conversely, whether its innovations are likely to cause harm long before that harm shows up in financial statements or sustainability disclosures.

Astellas Pharma illustrates a firm reinforcing an already strong sustainability profile. This Japanese pharmaceutical company focuses its research on oncology, urology, and rare diseases. The nearly 1,800 patents it filed over the past decade cluster in oncology biologics, antibody combination therapies, and gene-targeting approaches for conditions such as muscular dystrophy. These score highly in our model not only because they address serious unmet medical need, but because they represent novel therapeutic mechanisms rather than incremental reformulations of existing drugs. Notably, recent filings are more sustainability-positive than earlier ones – Astellas is not resting on a strong historical portfolio but actively building on it.

Solvay offers a different kind of signal. The Belgian speciality chemicals firm is not a sustainability leader today – its core business is basic and speciality chemicals. Yet its patent portfolio points in a different direction. Although the portfolio is small, the vast majority of filings signal sustainable inventions, anchored in two areas: waste purification processes and advanced electrochemical components, including battery materials and separators. Both map directly to the clean energy and circular economy agenda. Recent filings suggest the portfolio is becoming greener over time, not drifting back toward conventional chemistry – a quiet indication of where the company may be heading.

Not all firms, however, are innovating for sustainability. FMC Corporation, a global agricultural chemicals company, has filed more than 2,000 patents, yet nearly two-thirds scored negatively in our assessment. Its portfolio includes diamide insecticides, herbicidal compounds, and novel fungicides – chemicals that can meaningfully contribute to agricultural productivity by targeting pest species, but that are also expected to harm terrestrial and aquatic biodiversity. Pesticides and fungicides are among the leading drivers of biodiversity loss globally, and FMC's innovation pipeline points further in that direction.

Although patents do not always lead to improved operational processes or the delivery of novel products [118,119], they are a valuable early signal – one that revenue and CapEx data, by their nature, cannot provide. Together, sustainable revenue momentum, capital expenditure, and innovation pipelines give investors a richer, more temporally layered picture of where a company's sustainability impacts are heading. This is the fox's natural habitat: not a single metric, but a plurality of lenses, each revealing something the others cannot.

The next challenge is translating these insights into actual portfolios. How can investors construct sustainable portfolios and put these screens to work?

5.5 Portfolio Construction

Screening defines the universe, but investing is where the real work begins. At first glance, this may seem simple: allocate capital to companies with stronger sustainability performance and avoid the worst offenders, adjusting stringency as desired. In practice, it is more complex. Three dimensions warrant careful consideration: stringency, breadth, and financial performance.

5.5.1 Stringency

How sustainable do you want your portfolio to be? Do you build a 'dark green' portfolio that invests only in true leaders, or opt for a 'lighter green' fund that also includes neutral or even mildly negative firms?

A stringent approach has the benefit of delivering a greener portfolio. A less stringent strategy leaves more room for alpha, especially in the short term. The choice therefore depends on your other investment objectives: your performance target and your risk appetite.

5.5.2 Breadth

The next choice is about deciding whether your portfolio will remain agnostic to sustainability themes or whether it concentrates on specific sustainability challenges.

An agnostic portfolio offers broad thematic coverage. It may include companies advancing health, education, sanitation, climate action, biodiversity, and more – without prioritizing any single issue. This breadth allows for a wide-ranging impact.

A targeted portfolio narrows its focus to one or a few themes, seeking companies that actively contribute to solutions within those domains. But thematic focus must not compromise integrity. Staying true to the fox's plural philosophy, targeted investments should not undermine other sustainability objectives. For example, a climate-focused fund investing in a renewable energy firm that engages in corruption or mistreats workers risks greenwashing. A consistent approach would select a renewables company that advances climate goals without causing harm elsewhere.

In practice, the choice between agnostic and targeted portfolios often comes down to the investor's mandate and beneficiary preferences. Pension funds with broad societal responsibilities may favour agnostic approaches, reflecting the diversity of their members' values. Asset owners with a specific mission, like a pension fund for health workers, may opt for targeted strategies that concentrate impact where it matters most to them. Neither is inherently superior. What matters is that the chosen approach is applied consistently and transparently, and that thematic focus never becomes an excuse for ignoring harm elsewhere.

5.5.3 Financial Performance

Finally, sustainability does not replace financial fundamentals. Once the universe is screened, the standard investment process resumes: analyzing risk, return, valuation, and other metrics. A sustainable portfolio with weak financials will not endure. The challenge is to identify sound investments, and ideally those where sustainability acts as a source of alpha.

Beyond conventional analysis, investors should assess how the transition toward sustainability affects a company's financial profile. Some firms will invest in new business models that support future growth; others may incur costs without returns. Not all capital expenditures are equal: some are strategic, others – such as accounting for persistent externalities – are operational. In sectors undergoing deep transformation, investors must prepare for longer horizons and slower returns.

Consider, for instance, a chemical firm that invests heavily in pollution control – no small matter, given that chemical pollution ranks among the most urgent and underappreciated sustainability challenges of our time. If those capital expenditures merely limit negative externalities without changing the firm's product offering, they help maintain its licence to operate: a real strategic advantage, but one with modest return implications. Compare that to a firm investing in an entirely new, more sustainable product range. There, the same capital expenditure becomes a driver of future growth.

To navigate this distinction, investors must ask a pointed question of every sustainability-related investment: is this generating future value, or merely preserving today's licence to operate? When properly understood and integrated, sustainability becomes a risk factor – one that drives returns rather than compromises them.

5.6 Concluding Remarks: Invest Sustainably Like a Fox

The fox offers a clear advantage over the hedgehog in that it sees sustainability as it actually is: plural, interconnected, and full of trade-offs. A credible sustainability screen must reflect this complexity, acknowledging tensions across goals rather than collapsing them into a single metric or overarching lens.

This chapter has argued that effective sustainability screens must meet three criteria. They must assess companies across multiple sustainability goals simultaneously. They must deliver an overall judgement that prioritizes avoiding harm over rewarding positive contributions. And they must be grounded in sustainability science: evidence-based and adaptive, updating as knowledge advances rather than entrenched in outdated assumptions or emotive reasoning.

Meeting these criteria is harder than it sounds, and that difficulty explains why hedgehog approaches still dominate. Simplicity is cheap, but plurality is demanding. And the principal-agent dynamics between asset owners and asset managers create structural incentives that favour the hedgehog over the fox. Changing this requires transparency: asset owners must demand better screens, and asset managers must be willing to explain how their tools actually work.

SDG scores represent a meaningful step in the right direction. They typically are plural, harm-avoiding, and science-based. But they are backward-looking by design, capturing where companies are, not where they are going. The next generation of screens must add a forward-looking dimension, identifying companies that are credibly transitioning toward more sustainable business models.

Yet screening is only the beginning. To truly align capital with a sustainable future, investors must also consider portfolio construction: how green the portfolio should be, the breadth of sustainability themes it targets, and financial performance – ensuring that sustainability drives returns rather than comes at their expense.

With better, more holistic, and forward-looking screens, investors can build portfolios that are not only more sustainable but also more effective. But will better investing lead to a better world? The next chapter takes up this challenge.

6

The Impact of Sustainable Investing

Can Investors Save the World?

Sustainable investing is frequently championed as a tool for tackling the world's most pressing challenges. And many have called for the urgent deployment of private capital to help build a better future.

The launch of the Glasgow Financial Alliance for Net Zero (GFANZ) at the COP26 climate conference held in Glasgow in 2021 marked a high point in enthusiasm for the role of finance in advancing sustainable development. At the time, Boris Johnson, then Prime Minister of the United Kingdom declared: '*Uniting the world's banks and financial institutions behind the global transition to net zero is crucial to unlocking the finance we need to get there*'. John Kerry, the U.S. Special Presidential Envoy for Climate, added: '*The largest financial players in the world recognize energy transition represents a vast commercial opportunity as well as a planetary imperative*'. And Alok Sharma, the COP26 President-Designate, emphasized: '*Without adequate finance, we simply will not achieve the change needed to safeguard our planet for future generations. As the world continues down a crucial decade of delivery on climate action, GFANZ will ensure much-needed acceleration towards net zero by uniting some of the world's most powerful financial actors*' [120].

Yet this optimistic view is not universally shared. Critics have raised serious concerns about the ability of sustainable investing to deliver meaningful real-world impact.

In 2021, Tariq Fancy, the former chief of sustainable investing at the mammoth asset management firm BlackRock, famously described sustainable investing as '*a dangerous placebo that harms the public interest*'. He went further, likening it to '*selling wheatgrass to a cancer patient*' [121].

J. A. van Zanten, *Big Money, Better World?*,
https://doi.org/10.1007/978-3-032-32442-9_6

In his view, sustainable investing is a convenient fantasy that allows people to feel like they are making a difference, even when the actual impact is negligible. Worse, it may distract from the real solutions needed – like governmental regulation and systemic reforms that can really catalyze private sector action – which now are not receiving the attention they need. He warned that time is running out to act: '*we can no longer afford to answer inconvenient truths with convenient fantasies*'. As Fortune put it, Fancy '*more or less went nuclear, declaring the whole industry not just fuzzy and vague, but actively destructive*' [122].

So, which of these views is correct? Can sustainable investing help solve the world's sustainability challenges – or is it actually a harmful distraction from the real work that needs to be done?

Answering this question is far from straightforward. As we have seen, just as foxes employ an array of tactics, investors can pursue impact through different strategies. Two of the most prominent are asset allocation and active ownership – both of which can lead to *targeted* positive change. These can directly influence outcomes. A third pathway is through field building: changing the assumptions, norms, and rules of stakeholders that companies frequently interact with as a means of effecting *ecosystem-wide* changes [123]. These are indirect effects.

Yet even with these channels, the impact of sustainable investing is not immediate or guaranteed. Investors are often far removed from the real-world outcomes their strategies might influence. A common distinction is made between the *real economy* – where technologies are developed, infrastructure is built, food is produced, and goods are shipped – and the *financial economy* – where investors allocate capital to these activities. The chain of causality between investment decisions and real-world outcomes is long, diffuse, and often opaque.

So, let us trace this chain. Can sustainable investing truly shift the trajectory of global sustainability challenges – or is its impact more symbolic than substantive? In a world that is often fuzzy and vague, can we find clarity?

6.1 Asset Allocation

When investors use asset allocation as a strategy, they construct a portfolio that aligns with their sustainability objectives. There are many ways to do this. Some investors exclude companies with significant negative environmental or social impacts. Others apply best-in-class screening, allocating capital only to sustainability leaders – typically relative to their industry peers. Many strategies fall somewhere in between, such as underweighting 'brown' firms and overweighting 'green' ones. In all cases, the resulting portfolio is designed to be more sustainable than the broader market – and thus deviates from the so-called market portfolio.

How could deviating from the market portfolio contribute to positive change? Two mechanisms are relevant for us to investigate. First, it can create incentives for company management to improve sustainability performance. Second, it can influence a company's cost of capital. In short: the first mechanism affects the *quality* of a company's sustainability efforts, while the second affects the scale – or *quantity* – at which those efforts can be pursued [124].

6.1.1 Incentivizing Improved Sustainability Performance

Sustainable asset allocation can influence asset prices. For instance, when many investors avoid a company due to its adverse environmental or social impacts, its share price may decline. If company leadership perceives this undervaluation as linked to sustainability concerns, it may be motivated to improve its practices in order to regain investor confidence.

In 2020, a team of sustainable finance researchers at the University of Zurich reviewed the academic literature on the impact of sustainable investing. They reported evidence that it can affect asset prices but also noted a lack of consensus on the magnitude of this effect, and whether such price changes actually entice management to improve sustainability performance.

However, new studies are emerging rapidly to fill these gaps. For instance, a 2022 paper examined the 10% of mutual funds that are most aggressively reducing portfolio emissions by divesting from high-emitting firms [125]. It revealed that these 'decarbonization trades' lowered the stock prices of companies with high greenhouse gas emissions, which in turn incentivized them to reduce their emissions. Another 2022 study reached similar conclusions: it showed that companies with many sustainability-minded shareholders are fearful of divestment [126]. Even small divestments prompted these firms to improve their environmental and social policies to avoid further capital flight.

This growing body of research suggests that sustainable asset allocation can indeed incentivize companies to enhance their sustainability performance. As the field evolves, new studies will continue to shed light on the strength and scope of this relationship.

Moreover, the Zurich research team identified specific conditions under which asset allocation is more likely to incentivize companies to improve. They highlight three key factors.

The first is consistency in investor screening. The impact is stronger when a large share of investors uses similar criteria to assess sustainability performance. This consistency helps companies understand why they are being excluded or underweighted, enabling them to respond by improving in the areas where they fall short.

Second, the effect is likely to be greater for companies that have few close substitutes in the market. If investors can easily switch to another firm with similar financial characteristics – being correlated with the market portfolio – the price signal weakens. But when substitutes are scarce, the signal is stronger – making it more likely that management will take notice and act.

And third, the strategy is more effective when it is relatively easy or cost-efficient for companies to enhance their sustainability performance. In other words, firms are more likely to respond when the 'low-hanging fruit' is within reach – when meaningful improvements can be made without prohibitive costs or structural overhauls.

A compelling illustration of how these mechanisms can play out in practice comes from the classic business book *Barbarians at the Gate*, which chronicles the frenzied bidding war for control of RJR Nabisco – one of America's leading tobacco and food companies at the time, famous for brands like Camel and Winston and Salem cigarettes, and Oreos and Ritz Crackers.

After the October 1987 market crash, RJR Nabisco's share price plummeted. This marked the beginning of CEO Ross Johnson's downfall. The low price '*would haunt him for months to come*'. In December, the company boasted a 25% rise in profits, yet investors remained unmoved. Even as other food companies' share prices climbed, RJR Nabisco's stock stagnated.

A year later, Johnson lamented the persistent undervaluation: '*It's plain as the nose on your face that this company is wildly undervalued… We tried to put food and tobacco businesses together, and it hasn't worked. Diversification is not working. We are sitting on food assets that are worth twenty-two, twenty-five times earnings and we trade at nine times earnings, because we're still seen as a tobacco company*'. As the book explains, Johnson, like many other executives, viewed the stock price of his firm like a report card and '*as he watched other food stocks soar, Johnson felt like a wallflower at the orgy*'.

Johnson tried to boost the share price beyond diversifying into food. One gambit, dubbed Premier, aimed to develop a high-tech 'smokeless' cigarette with minimal smoke and no tar. It was trumpeted as the company's '*secret weapon to turn the tide on the antismoking movement*', but the project flopped and was widely ridiculed. Ultimately, when all else failed, Johnson pursued a leveraged buyout to address the low stock price, triggering the dramatic bidding war that the book narrates like a financial thriller.

Although it is a unique case, it illustrates how sustained investor aversion can push management towards increasingly bold (and sometimes desperate) strategies. In line with the first condition discussed earlier, Johnson interpreted investor aversion to tobacco as the root cause of the company's undervaluation. The stigma – and the depressed share price – extended to other tobacco firms as well, suggesting a lack of close substitutes, consistent with the second condition. However, the third condition proved elusive: making a tobacco company more sustainable was not easy. Diversification failed to shake the stigma, and the costly attempt to develop a cleaner cigarette fell flat. Still, the case shows that investors can create incentives through asset allocation – motivating companies to change, even if the path to sustainability is far from straightforward.

6.1.2 Affecting the Scale of (Un)Sustainable Activities

Beyond incentivizing companies to change, asset allocation can also influence the scale of corporate operations by altering the cost at which they access capital.

As the Zurich research team explains, investors can achieve this through both direct and indirect channels. The direct channel involves investors financing a sustainable company on better terms than the company would get from other investors with neutral sustainability preferences – much like a mortgage provider giving a discount to borrowers buying a more energy-efficient house. This approach is commonly used by development finance institutions and other public sector actors.

Increasingly, private investors are also adopting tools to influence capital costs. One such instrument is the sustainability-linked bond (SLB). What sets SLBs apart is that the coupon rate – the annual interest paid by the issuer to the bondholder – is tied to the achievement of specific sustainability targets. Hence, unlike green bonds, which earmark proceeds for specific environmental projects, SLBs link financial terms to company-wide sustainability performance.

For example, in 2020, the European cement company Holcim issued an €850 million SLB with a coupon of 50 basis points, maturing in 2031. If the company failed to meet its 2030 greenhouse gas emission reduction target, the coupon would increase to 75 basis points.[1]

[1] The company's 2030 greenhouse gas emission reduction target is to emit 475 kg of CO_2 emissions per ton of cementitious material produced. In 2019, this value stood at 561 kg.

This structure creates a clear financial incentive: achieving sustainability goals results in a lower cost of capital, while falling short leads to higher costs. The effectiveness of SLBs in reducing capital costs is supported by a broader study comparing SLBs to conventional bonds across a wide sample of companies. The study found that SLBs carry a premium of -21.5 basis points, meaning that, on average, companies issuing SLBs pay 0.21% less than they would for a plain vanilla bond [127].[2]

While SLBs are promising tools for investors seeking to lower the cost of capital for sustainable companies, they currently represent only a small (though growing) share of the global bond market. Moreover, they are not available to equity investors.

A second, more indirect, channel through which investors can influence the cost of capital, relates to the influence of sustainable investing on asset prices. As discussed above, this may not only be an incentive for management to improve sustainability performance, but it can also change the cost at which affected companies raise capital from other investors.

Suppose, for instance, that an investor divests from an unsustainable firm – say, a coal mining company. Concerned about the company's high greenhouse gas emissions and seeing no viable path to influence its practices via active ownership, the investor pulls her money out. That money needs to go somewhere – otherwise it sits idle in the portfolio, yielding limited return. To avoid this, let us imagine she reallocates it to a renewables company.

Now what happens because of this divestment and reinvestment?
One possible effect is a shift in the cost of capital for both firms. If many investors divest from the coal company, the price of its stock (or bonds) may fall, making it more expensive for the company to raise funds through equity or debt in the broader market. Consequently, this makes it more expensive for the firm to operate, and expansion of its unsustainable business activities may become less financially viable – a clear win for climate change.

Conversely, increased investment in the renewables company could raise its equity or debt price and lower its cost of capital, enabling it to expand more easily and make a greater impact at a lower cost.

[2] Despite these positive findings in the extant literature, SLBs deserve scrutiny. While some sustainability-linked bonds are credible tools that lower capital costs for issuers making sustainability progress, many are too light to matter. Two recurring issues stand out. First, weak targets: issuers can secure a lower coupon despite minimal progress by anchoring their KPI to a far-back baseline year or otherwise unambitious levels. Second, duration: some SLBs have a target measurement date that is set only a few years prior to the bond maturing, leaving only a few coupons exposed to a potential penalty step-up. And, in some cases, the bonds are callable before the measurement date, giving the issuer the option to redeem early and avoid the KPI measurement entirely. These features create room for greenwashing and warrant careful investor scrutiny before allocating capital.

This mechanism is theoretically compelling, but empirically hard to examine. The reality is that the impact of sustainable investing on a company's cost of capital is not always direct or immediate. Its effectiveness depends on scale, context, and market dynamics.

Let us return to our investor. When she sells her stake in the coal company, someone else buys. The asset simply changes hands – like a baton passed in a relay race. The direct impact of this trade on the company's operations is not immediately obvious.

In the *Solving for Climate Podcast*, Alex Edmans – the finance professor at London Business School, who we met in Chapter 2 – offers a helpful analogy. He contrasts divestment with consumer boycotts, using the example of McDonald's. A boycott will lead to hamburgers sitting on the shelves unsold, dropping revenues, and putting pressure on McDonald's. But, he explains, '*sadly that is not the same with our money. Why? Because if I own some stocks and if I was to sell my shares … I can't sell unless someone else buys*' [128].

Yet does divestment have no effect at all? Or might there still be some influence on a firm's cost of capital?

This question is hotly debated by academics. Increasingly, a body of research is emerging that sustainable investing can result in measurable, but modest, influence on cost of capital.

An important paper, entitled '*The Impact of Green Investors on Stock Prices*' [129], features a theoretical model in which green investors avoid investing in brown companies. This lowers the share price of those firms in order to attract other investors to buy them. Consequently, the cost of capital of brown firms increases. The authors then make assumptions about the proportion of sustainable investors in the market and divestment timeframes. Based on this, they estimate that divestment has a modest effect on the cost of capital for brown firms – increasing it by around 20 basis points.

Not all models agree, however, and the divergence is worth exploring. Another influential study – '*The Impact of Impact Investing*' – reaches a strikingly different conclusion [130]. It estimates that, in order for sustainable investing to affect a company's cost of capital by more than 1%, it would need to represent 80% of all investable wealth. Based on the authors' assumptions and analysis of U.S. stock markets, the actual effect is estimated at just half a basis point – a change considered economically insignificant.

What explains the divergent findings of these two papers? The first model assumes that passive investors – who simply track a broader index – will not absorb the brown stocks that sustainable investors sell, leaving a smaller pool of potential buyers and therefore a stronger price signal. The second model assumes a larger and more fluid pool of substitute buyers, which dilutes the signal considerably. What this means in practice is that the effect of sustainable investing on cost of capital is real but uncertain in magnitude – and likely depends heavily on how concentrated and coordinated sustainable investors actually are in any given market.

In any case, models only take us so far in understanding how sustainable investing might move the cost of capital [131]. Even if green activities face lower financing costs (and brown activities higher), that relationship matters only if companies themselves feel it – that is, if corporate managers perceive green projects as cheaper to finance than brown ones. If they do not, the mechanism exists on paper but has negligible real-world impact.

Exactly this perception channel is tested in a recent working paper [131]. The authors analyze conference-call transcripts from 730 large, high-emitting companies, focusing on how cost of capital and discount rates are discussed. This gave the authors an insight into how companies perceive their cost of capital over time, which they could combine with how green the firms were.

The results are striking. As sustainable investing accelerated after 2016,[3] the perceived cost of capital for greener firms fell substantially relative to brown ones. Prior to 2016, both groups looked similar. From 2016 to 2023, by contrast, the gap started to widen. On average in this period, green firms enjoyed a 1 percentage point lower perceived cost of capital than brown firms.[4]

This is a substantial effect – supporting the view that sustainable investing can influence companies' financing incentives, nudging capital formation toward greener activities.

To take a final look at this relationship, we can turn to green bonds. In fixed income markets, cost of capital is easier to observe than in equity markets – thanks to the transparency of coupon rates and related yield to investors. Green bonds offer an additional advantage: their proceeds are earmarked

[3] The rise of sustainable investing has surged since 2016. Some relevant drivers were the adoption of the Sustainable Development Goals (SDGs) and the Paris Agreement in late 2015, both of which drew attention to the role of investors in attaining global policy goals for sustainable development.

[4] This discussion compares green firms to brown ones. In addition to these cross-firm effects, the researchers also found a within-company effect. Some sophisticated companies – like BP, EDF, Shell, or TotalEnergies – started to use lower cost of capital and discount rates for their renewable energy divisions relative to their fossil fuel divisions. Their greener divisions enjoyed on an average, one percentage point lower cost of capital – consistent with the cross-firm estimate explained above – making the financing of green activities more attractive.

for environmentally sustainable activities, making it clearer to investors what their capital is funding. In this way, green bonds help overcome two major hurdles in empirically assessing the link between sustainable investing and cost of capital.

A recent study assessed the borrowing cost advantage that green bond issuers enjoy at the time of issuance [132]. It examined 1,169 green bonds and 129,043 conventional corporate bonds issued between 2014 and 2021, across nearly 13,000 firms worldwide. The paper found that, on average, green bonds offer a direct borrowing cost advantage of between 3 and 8 basis points. This translates into a reduction in borrowing costs of between 2% and 7% – a modest but meaningful difference.[5]

Where does this leave us?

Sustainable investing seems to have a small but meaningful effect on the cost of capital. It can reward sustainable, and penalize unsustainable, activities. This can scale sustainable activities – because they become more affordable – and make unsustainable more costly and therefore less attractive to pursue. This link between the cost of capital and sustainability performance is not only theoretical. Recent research finds a plausibly causal effect: companies that face lower financing costs due to sustainable investing measurably improve their sustainability performance as a result [133].

That said, the relationship is complex and perspectives diverge.

Equally clear, however, is that two key conditions increase the likelihood that investors can meaningfully influence capital costs. First, there must be a broad consensus on which companies to support or avoid. If investors can agree on this, more capital will flow towards sustainability leaders and less to laggards, adding pressure on companies and their cost of capital. The current state of uncorrelated sustainability screenings in the market is not helpful towards this objective. Second, sustainable asset allocation decisions must be backed by substantial financial weight. Only then would the remaining buyers be forced to hold more concentrated portfolios – taking on more risk and demanding higher returns. That is when prices would shift meaningfully, and the cost of capital would rise in a way that could influence corporate behaviour.

In sum, sustainable investing can influence the cost of capital, especially when it gathers financial weight.

[5] As the authors of the study caution, this advantage may be offset by the fees and compliance costs associated with issuing green bonds.

6.2 Active Ownership

If asset allocation is the investor's exit, active ownership is their voice – and it can be surprisingly powerful. Through proxy voting and engagement, investors can push companies to change from the inside rather than withdrawing capital altogether.

Proxy voting can be an effective tool. While research shows that most sustainability-related proposals are still rejected – particularly by large and passive investors – approval rates have been rising over time [56]. Companies tend to be more responsive to proposals submitted by institutional shareholders with a long-term perspective, those that can influence corporate reputation, or those located geographically close to the firm [123]. Yet, despite these promising trends, the evidence remains inconclusive. It is still unclear whether proxy voting alone leads to measurable real-world impact.

Engagement activities have received more attention from researchers and practitioners alike. These ongoing dialogues offer a more sustained and targeted form of influence compared to the episodic nature of proxy voting. Moreover, studying the impact of proxy voting is complicated by the fact that some politically conservative investors submit sustainability-related proposals not to encourage progress, but to resist it. As a result, some proposals labelled as sustainability-related in fact push in the opposite direction, making it hard to tell whether support for a proposal really advances sustainability. Engagement, by comparison, often provides clearer pathways to observe positive change.

Many investors report on their engagement activities, outlining their objectives and claiming success. Self-reported data has an obvious limitation, however. Asset managers have an incentive to present outcomes favourably. Nonetheless, they are worth examining.

For instance, in 2024 my former employer Robeco engaged with eleven financial institutions on climate and nature transition topics, and with eight palm oil producers – reporting that over 70% of these engagements were either successful or showed positive progress [134]. Other topics present a more mixed picture. In the case of 'Net Zero Carbon Emissions', Robeco targeted 49 companies, with approximately 60% showing flat progress, nearly 30% demonstrating success or positive momentum, and just over 10% showing negative or no progress.

Robeco carefully selects the companies it engages with and follows a pragmatic approach, ideally encouraging changes that also enhance long-term financial performance. This may partly explain the relatively positive outcomes. Is there more robust, independent evidence on the actual impact of engagement?

The earlier-discussed review paper of the team from the University of Zurich also examined the evidence on the impact of engagement. It assessed five studies that evaluated the success rate of various engagement efforts. While engagement is not always successful, the studies found success rates ranging from 18% to 60% – suggesting a reasonable probability that engagement can lead to meaningful change.[6] Altogether, the review paper concludes that '*shareholder engagement is an effective mechanism through which investors can trigger reforms that improve the quality of company activities*'.

A recent study by sustainable finance professors Florian Heeb and Julian Kölbel – both part of the Zurich team mentioned earlier – adds further weight to the evidence that engagement can drive corporate change [135]. What makes their study particularly compelling is its experimental design, which allows for a more rigorous assessment of causality – something that is often lacking in this field.

They examined over 1,200 companies, randomly selecting 300 to receive a letter from an index provider. The letter encouraged these companies to reduce their greenhouse gas emissions and specifically requested that they commit to setting a science-based climate target – a condition for remaining included in the provider's climate transition benchmark indices.

One year later, the researchers assessed the outcomes. They found that 21% of the companies that received the letter had made a commitment, compared to just 15.7% in the control group that received no communication. The study concluded that this suggests '*engagement by financial institutions can affect corporate policies when a feasible request is combined with a credible threat of exit*'.

I find this study particularly insightful because it not only demonstrates a measurable effect but also highlights the conditions under which engagement is more likely to succeed.

One such condition is that the request must be feasible and not overly costly. Setting a science-based emissions target is within reach for most firms – it is not a moonshot. Another is that the investor must have influence, whether through significant ownership (individually or in coalition with others) or through social and cultural proximity to the company. A company's prior

[6] Importantly, these rates are not solely based on self-reported outcomes. Two of the studies, for instance, provided evidence that the ESG ratings of targeted companies improved over time, indicating that engagement can result in observable changes in company behaviour, as reflected in third-party data.

experience with ESG-related initiatives matters: firms with more experience are more likely to respond positively than those less familiar with sustainability integration. And finally, if there are consequences to non-compliance, then that helps companies to move, too [135–137].

6.3 Field Building

A team led by my coauthor Emilio Marti argues that investors often have a too narrow view of how they can create impact. While asset allocation and engagement – exit and voice – are important, they are not the only strategies available. Drawing on extensive literature and conversations with investors, the team identifies a third strategy: *field building* [123].

This strategy starts from the observation that companies do not act in isolation. They are embedded in 'fields': the networks and environments in which firms operate, including industries and supply chains, but also the stakeholders they interact with most frequently, such as investors, regulators, clients, and the media. These fields shape what companies view as legitimate and influence their behaviour.

A personal example illustrates what this means in practice. While at Robeco, I helped launch the Open Access initiative, through which Robeco made its SDG scores for around 12,000 companies publicly available via its website and through Bloomberg. The logic was simple: if we wanted to maximize our impact, we needed to share our intellectual property. While protecting it might make commercial sense, sharing it would help reduce confusion around sustainability ratings and mobilize additional capital toward sustainability leaders – and away from laggards. This is field building: not changing a single company's behaviour directly, but reshaping the information environment in which thousands of companies and investors operate.

Emilio and his team argue that investors can advance sustainability by reshaping these fields more broadly: influencing the norms, assumptions, incentives, and rules that surround companies. This is a wider toolkit than most investors use, and the evidence suggests it can be powerful. The team identifies several concrete tactics.

Some of these tactics aim to influence companies indirectly by targeting other shareholders. One approach is to *shift how other shareholders evaluate sustainability issues*. Through divestment campaigns, investors can create normative pressure that encourages others to sell a company's stock – an approach that has proven effective in various cases.

Another tactic is *sharing expertise*. Sustainable investing is marked by significant information asymmetries: some investors possess deep knowledge of sustainability, while others do not. Sharing this expertise – whether through open data initiatives, collaborative research, or industry forums – can mobilize and motivate others to invest more sustainably. The Robeco Open Access initiative, discussed earlier, is one example of this in practice [138].

Other tactics focus on shaping the broader context in which companies operate.

One such approach is to *delegitimize harmful business activities*. By investing sustainably – and especially through divestment, shareholder proposals, and public communication – investors can visibly signal their values. This helps stigmatize unsustainable practices and elevate sustainability goals, nudging companies to align their activities accordingly.

For example, when Norges Bank Investment Management (NBIM) divested from Caterpillar over concerns about its alleged complicity in Israel's war on Palestine, it sparked global media attention. Other investors, including Dutch pension giant ABP, followed suit in excluding Caterpillar. When influential investors like NBIM and ABP publicly state that certain behaviours are unacceptable, they help shape norms and values across the broader investment community.

Another tactic is the *establishment of voluntary standards*. Investors have frequently collaborated to shape sustainability-related standards, such as the Global Reporting Initiative, which standardizes corporate sustainability reporting, and the Taskforce on Nature-Related Financial Disclosures, which sets voluntary guidelines for how companies report on their environmental impacts and nature-related risks.

Finally, investors can *support regulatory change*. Once a critical mass of companies voluntarily adopts sustainability practices – demonstrating their technical and financial viability – regulators may step in to standardize these practices. Investors can play a key role in shaping such regulations. Given their exposure to a wide cross-section of industries, they possess valuable insights that can inform policymakers on how to advance sustainability effectively.

Taken together, these tactics show that investors are not merely market participants – they are field shapers. By expanding their toolkit beyond exit and voice, they can help redefine the rules of the game and accelerate the transition to a more sustainable economy.

6.4 Concluding Remarks: Neither Placebo Nor Panacea

This chapter opened with high expectations about the contribution that sustainable investing could make in addressing global sustainability challenges – followed by sharp criticism suggesting it distracts from real solutions, serving as a *'dangerous placebo'.*

Like a fox, sustainable investors can employ many tactics. There is modest but relevant evidence that sustainable investing can incentivize companies to become more sustainable by influencing asset prices, that it can raise the cost of capital for unsustainable firms while lowering that of sustainable ones, and that active ownership could directly induce positive behaviour. More broadly, they can shape the fields in which companies operate, with growing evidence that investors can shift norms and values through their non-market behaviours.

What is also clear is that scale matters. Niche investment strategies are unlikely to move markets or meaningfully influence corporate behaviour. Sustainable investing is most effective when it competes with mainstream portfolios. Another critical factor is a shared understanding of corporate sustainability performance. When investors use divergent screening methods – such as ESG ratings that lack correlation – the signal to companies weakens, reducing the impact on managerial incentives and cost of capital. Shared sustainability screens, even if only at the extremes of highly (un)sustainable companies, can serve as a guiding light toward greater impact.

This evidence does not make sustainable investing a cure-all. There are pathways for it to generate positive impact, but it will not drive systemic change on its own. That still requires government intervention to reshape consumption and production patterns. The good news is that sustainable investing does not appear to distract from other solutions, as the 'dangerous placebo' analogy implied. There is no evidence that it crowds out support for sustainability policy [139].

I interpret the evidence modestly: sustainable investing is a relevant instrument that investors can wield to promote sustainable development – but it is not a silver bullet. This modesty stands in contrast to the bold marketing claims often made by financial institutions.

Sustainable investing is relatively simple in the grand scheme of solving global problems. To invest sustainably, you do not need to be a genius. You are not inventing clean energy systems or developing novel medicines. Instead, you are screening companies on their sustainability performance, allocating capital accordingly, applying active ownership, and shaping fields. This will not revolutionize the world or eliminate the root causes of sustainability challenges.

But that does not make it unimportant. Although governments must lead, the private sector is also required – and expected – to play a role.

A foxy investor does not ask simply whether sustainable investing has impact. Rather, they ask which channel, under which conditions, produces the most impact. The evidence here is instructive. Asset allocation works best when many investors screen consistently, and companies have few substitutes. Engagement succeeds when requests are feasible and consequences are credible. And field building operates at a different scale altogether, reshaping the context in which thousands of companies make decisions simultaneously. Together, these three channels are more powerful than any one alone.

Wangari Maathai, Nobel Peace Prize laureate and founder of the Green Belt Movement, captured the power of collective action well: '*Everyone of us can make a contribution. And quite often we are looking for the big things and forget that, wherever we are, we can make a contribution. Sometimes I tell myself, I may only be planting a tree here, but just imagine what's happening if there are billions of people out there doing something. Just imagine the power of what we can do*' [140].

Sustainable investing will not save the world on its own. But that is the wrong standard. The right question is not whether it solves everything, but whether it does something – and whether that something is worth doing. The evidence suggests it is. The fox's task is not to pretend otherwise, but to be honest about the magnitude, rigorous about the mechanisms, and ambitious about the direction. That combination of honesty, rigour, and ambition is what separates impact from illusion.

7

Sustainability and Financial Performance

Will Sustainable Investing Create Wealth or Wipe It Out?

It should be illegal for investors to bother with sustainability. At least, that is what some people think. Florida's governor, Ron DeSantis, has likened sustainable investing to '*ideological joyrides*' [141]. His stance is part of a broader political campaign against what he refers to as '*woke capitalism*'. Former U.S. Vice-President Mike Pence claimed that sustainable investing '*empowers an unelected cabal of bureaucrats… to rate companies based on their adherence to left-wing values*'. As a result, sustainable investing has become entangled in America's 'Culture Wars'. In several Republican-led states, legislation has been passed or proposed to ban the integration of sustainability factors into investment decisions. In Florida, for example, state and local governments are prohibited from considering sustainability in how they invest public funds [142].

Beyond the punchlines, the argument is straightforward: sustainability distracts investors from their core purpose of maximizing returns. The logic goes like this. If you are an investor, your job is to manage money and to grow that money while controlling for financial risks. Anything else, especially sustainability, is a distraction that should be outlawed.

But turning sustainable investors into villains is short-sighted and self-serving.

For one, it simply is impossible to separate sustainability from investing. Sustainability can influence investment results. Poor sustainability performance may create material risks that lead to losses. Strong sustainability performance could drive financial results. Rather than harming returns, sustainability may enhance them. Evidence is needed to understand how these dynamics play out in practice – and that is exactly what we will explore.

J. A. van Zanten, *Big Money, Better World?*,
https://doi.org/10.1007/978-3-032-32442-9_7

This conclusion is echoed in studies investigating if better sustainability ratings are associated with stronger financial performance.

One particularly comprehensive analysis explored the relationship between ESG ratings and stock returns across more than 16,000 stocks in 48 countries, over a span of 20 years [143]. Using seven distinct ESG ratings, the researchers found very little evidence that ESG ratings were meaningfully related to stock returns — positively or negatively. In short, ESG ratings have not significantly influenced investment performance over the past two decades.

For some advocates of sustainable investing, this outcome may be disappointing. If sustainable companies consistently outperformed, it would offer a compelling argument for scaling up sustainable investment strategies. Conversely, critics might seize on these findings to argue that sustainable investing offers no financial benefit and is therefore a costly distraction.

Yet there is also a more optimistic interpretation. If sustainable investing does not compromise financial performance, then investors can build sustainable portfolios without sacrificing returns. In effect, sustainability comes at no extra cost.

It is also worth remembering that sustainability is a broad and multifaceted concept. The studies reviewed span a wide range of sustainability issues and investment styles – from divestment and active ownership to ESG integration and carbon emission reductions. Some sustainability topics and some investment styles may be more closely linked to financial performance than others. This suggests that while general conclusions may be mixed, there are still opportunities to generate outperformance through sustainability. The challenge lies in knowing where to look for the premium and how to capture it.

7.2 The Investment Implications of Sustainable Investing

Yet even if a sustainability premium exists – say, when green companies outperform or brown firms underperform – it does not automatically mean that investors will benefit from it.

Imagine you are an investor who has screened companies in your investment universe based on their sustainability performance. You have tested this screening method and found that the more sustainable companies tend to deliver stronger returns. Encouraged by this result, you build a portfolio that invests in the sustainability leaders and excludes the laggards.

Over time, you track the portfolio's performance. But to your disappointment, it does not show the expected outperformance. What might be happening?

By excluding the sustainability laggards, your portfolio may have lost some of its diversification. This can expose it to higher risk and potential underperformance. Additionally, your portfolio might have weaker exposure to certain factors – characteristics of companies that have historically been linked to improved financial performance, such as size, value, or momentum. Missing out on these factor premiums could further drag down returns.

In other words, even if a sustainability premium exists at the company level, it may not be sufficient to offset the loss of diversification and factor exposure at the portfolio level. Moving from theoretical insights to actual investment practice introduces complexity – and potential trade-offs – that investors need to navigate carefully.

This shift from theory to practice is critical. A theoretical link between corporate sustainability and financial performance only matters if it can be captured in actual investment portfolios. That is why, together with my colleagues Joop Huij – who designed the study – and Dries Laurs – who led the data analysis – we set out to test what happens to sustainable investment portfolios in practice [145].

We began by constructing portfolios based on sustainability criteria. Recognizing that there are multiple ways to screen companies – and setting aside, for now, the question of how accurate these measures are - we used three common approaches: SDG scores, ESG ratings, and carbon emissions data. To build the portfolios, we applied increasingly stringent exclusions, investing in better-performing companies and avoiding those with weaker sustainability profiles. In our main analyses, we excluded roughly a quarter of the companies in the investment universe. This left us with portfolios consisting of about 75% of available companies, though the specific companies included varied depending on the screening method used.

Next, we examined the implications from an investment perspective. Specifically, we focused on two key objectives that sustainable investing could affect: diversification benefits and factor exposures.

7.2.1 Diversification Benefits

Diversification is one of the most fundamental principles in investing. The idea is simple: by spreading investments across a broad range of companies, you reduce the risk that any single poor performer will drag down your entire portfolio. If one company faces headwinds, another might benefit from tailwinds that help offset the loss. Markets are unpredictable, and even well-managed firms can stumble. Diversification helps cushion against these shocks, offering more stable and resilient performance.

But diversification is not just about avoiding losses – it is also about capturing opportunities. Different sectors and regions perform well at different times, and a diversified portfolio is better positioned to benefit from these shifts.

Yet when investors narrow their focus too much – say, by excluding entire categories of companies based on sustainability criteria – they may unintentionally concentrate their risk. For example, avoiding fossil fuel companies could cause a portfolio to lag during an energy rally. This can lead to greater volatility and, in some cases, lower returns.

One way to measure the benefits of diversification is through the concept of the *efficient frontier*. This idea traces back to the foundational work of Harry Markowitz, who revolutionized investment theory in the 1950s. The efficient frontier is a curve that maps out the best possible combinations of risk and return that a portfolio can achieve. Markowitz's insight was simple yet profound: investors should not only consider the expected return of individual assets, but also how those assets interact within a portfolio. By analyzing correlations and optimizing combinations, he demonstrated that a well-diversified portfolio could deliver superior risk-adjusted returns compared to any single investment. His work became a cornerstone of modern finance and ultimately earned him the Nobel Prize in Economics. Remarkably, when defending his PhD thesis in 1954, the renowned economist Milton Friedman – who also has a Nobel – challenged his ideas, arguing they were not economics [146] – a testament to how far ahead of his time Markowitz truly was. Today, the efficient frontier remains a powerful tool for understanding how investors can balance risk and return – even when sustainability is introduced as an additional objective.

Portfolios that lie on the efficient frontier are considered optimally diversified: they offer the highest expected return for a given level of risk, or the lowest risk for a given level of return. If a portfolio is poorly diversified, it may fall below this frontier, meaning it is taking on more risk than necessary or missing out on potential returns.

We crunched the numbers to calculate the efficient frontier for our three sustainable investing portfolios – based on SDG scores, ESG ratings, and carbon emissions – and compared them to the efficient frontier of an unrestricted investment universe. To ensure robustness, we used a long timeframe: January 1985 to May 2021.

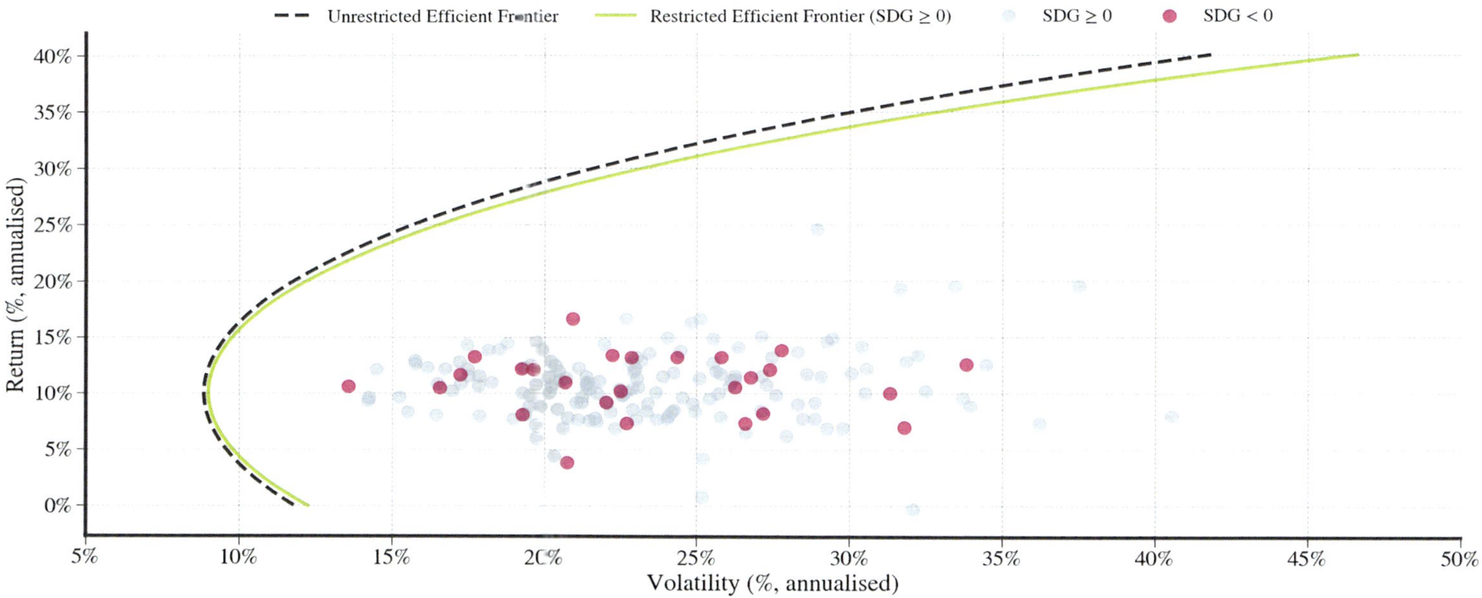

Fig. 7.1 The efficient frontier of SDG-aligned investment strategies. This figure shows the efficient frontiers of two portfolios. The 'Unrestricted Efficient Frontier' in black represents the market portfolio, while the 'Restricted Efficient Frontier' in green is an SDG-aligned portfolio that excludes investments that have negative impacts. The dots represent individual investments, whereby those in red have a negative impact, and those in blue are neutral or positive for the SDGs. Source: Huij, J., Laurs, D., & van Zanten, J. A. (2025). The investment implications of sustainable investing. *Journal of International Money and Finance*, 103372.

The findings are clear. The efficient frontiers of the sustainable portfolios overlap with those of the unrestricted universe – illustrated in Fig. 7.1, which compares an unrestricted portfolio with an SDG-aligned one that excludes around 25% of the market. Over the long run, sustainable investing does not appear to compromise diversification. And this result holds across all three screening methods.

7.2.2 Factor Exposure

In investment theory and practice, *factors* are characteristics of companies that help explain differences in their financial performance.

In the early 1990s, Eugene Fama and Kenneth French introduced the *three-factor model*. It showed that market risk alone could not fully explain asset returns. They added size (small versus large companies) and value (cheap versus expensive stocks) as additional factors, providing a more nuanced framework for portfolio construction and performance attribution. Much like Markowitz's efficient frontier, factor models help investors understand the trade-offs between risk and return. The groundbreaking work of Fama and French not only led them to be awarded Nobel prizes, but it also sparked factor investing – an approach that seeks to systematically capture these drivers of return by building portfolios that aim to outperform the market or manage risk more effectively.

Over time, the list of recognized factors has expanded. Some of the most widely studied and applied factors include value, momentum, quality, and low-risk. These are not abstract concepts – they are grounded in decades of empirical research and are used by asset managers around the world.

Value refers to companies that are priced cheaply relative to fundamentals like earnings or book value. These firms may be overlooked or temporarily out of favour, but historically they have tended to deliver strong returns over time. Momentum captures the tendency of stocks that have performed well recently to continue performing well in the near future. Quality focuses on financially healthy companies – those that are safe, profitable, expanding, and well-managed. And low-risk (or low volatility) targets companies that tend to fluctuate less in price, offering more stable returns. Each of these factors represents a different way of tilting a portfolio, and they can significantly influence both risk and return outcomes.

In our paper, Joop, Dries, and I tested a common concern: that tilting a portfolio toward sustainability leaders might cause investors to miss out on factor premiums – the excess, long-term returns that can be earned by investing in companies with good factor exposure. To explore this, we devised simple investment strategies designed to capture each of the four factor premiums. We then calculated the factor premiums for our sustainable portfolios, as well as for an unrestricted portfolio.

Our findings showed that across all four factors, the premiums remained largely consistent between the sustainable and unrestricted portfolios. We concluded that '*the long-term returns of factor strategies have not been affected by implementing sustainable investing*'. In other words, investors can pursue sustainability goals without sacrificing exposure to key drivers of financial performance.

7.2.3 Fear of Missing Out

Our investigation showed that sustainable investing does not need to compromise key investment objectives, such as achieving diversification or capturing factor premiums. Importantly, these results hold over the long run – even though year-to-year outcomes may vary.

Around the time our study was published, another research team at Erasmus University investigated a similar question using a different approach. Their analysis focused specifically on concentration risk: what happens when portfolios become less diversified, potentially as a result of sustainability-driven exclusions. Like us, they examined a long-time horizon, starting in 1985 and ending in 2023.

Their findings complement ours. They confirm that constructing a well-diversified portfolio typically requires more than 30 to 40 stocks, but not many thousands. They then go further, exploring the risks of under-diversification. One striking insight is that just 2% of stocks account for all net wealth creation in the stock market. The recent dominance of the so-called magnificent seven – Apple, Microsoft, Alphabet, Amazon, Nvidia, Meta, and Tesla – illustrates this point. These companies have driven a disproportionate share of market gains in recent years.

This concentration creates FOMO: a fear of missing out. If investors exclude many of these top performers – perhaps due to concerns about their sustainability impacts – they may inadvertently drag down portfolio performance. In other words, sustainable investing must be carefully balanced to avoid missing the small fraction of companies that drive the global equity premium.

This is where the fox's plural approach offers a practical advantage. Whereas the hedgehog's singular focus might rigorously cut companies from the investment universe, the fox's plural screen evaluates firms across multiple goals, reserving stringency for true laggards rather than applying it as a blunt instrument.

7.3 Other Benefits

The evidence so far suggests that sustainable investing comes at no cost in the long term. That sounds appealing – sustainability seemingly for free – but some critics lament the lack of a solid business case. If there is no financial upside, why bother?

But by focusing only on risk and return, are we not taking an overly narrow, hedgehog-like view? Direct financial benefits may not be the only reason to invest sustainably. Behavioural finance reminds us there are many valid motivations beyond performance. In his excellent book *Finance for Normal People: How Investors and Markets Behave* [147], Professor Meir Statman offers a striking analogy:

Imagine you want to give your beloved a rose. Your Finance 101 class taught you the rose has no utilitarian benefit and will wither in days. Would it not be better to gift $10 instead – the price of the rose – so your beloved can maximize their 'utility function?' This might be rational, but it is rather silly. Roses carry expressive and emotional value: saying 'I love you' in a way cash never can.

Sustainable investing works much the same way. Beyond utilitarian benefits like risk and return, it offers expressive benefits like signalling social responsibility and emotional benefits related to feeling good about aligning money with values. These may be harder to quantify, but they matter. Let us explore them.

7.3.1 Emotional Benefits: Aligning Money and Morals

The main emotional benefit of sustainable investing is simple: it lets money reflect moral values. For many, this alignment matters deeply. A health professional may not want their pension invested in tobacco companies that fuel the very illnesses they work to cure. A teacher might prefer their savings to support education rather than firms that exploit child labour. As we have seen earlier, research confirms this is not a niche concern: many people would rather see their pension aligned with sustainability goals, even if it means sacrificing some financial performance.

But sustainable investing is not only about funding companies you like. It can also make those values actionable. When investors allocate assets based on sustainability criteria, they naturally exclude firms with significant negative impacts. This leads to smaller, more focused portfolios. Instead of owning the entire market – say, 10,000 stocks – an investor might hold around 1,000 companies selected for both financial prospects and sustainability performance. With fewer holdings, each stake becomes relatively larger, increasing influence.

That matters. Proxy voting carries more weight when an investor owns a meaningful share. Engagement efforts – such as requesting meetings or proposing changes – are more likely to be taken seriously. Simply put, companies listen more closely to investors who hold a bigger piece of the pie. This increased influence means sustainable investing does more than align capital with moral values: it actively works to implement them.

What is more, there is also a financial case for this increased influence. Research finds that when engagement dialogues are successfully concluded on material sustainability issues, targeted companies outperform their peers by as much as 2.5% over the following 14 months [148]. Other studies show that firms subject to successful engagements outperform their non-targeted peers by 2.7% over the subsequent six months, with the effect rising to 7.5% for companies that initially had poor sustainability profiles [149]. Sustainable investing, in other words, does not just align money with values; it also supports the addition of financial value.

7.3.2 Expressive Benefits: Enhancing Reputations

Many investors care deeply about their reputation – especially institutions with a strong social profile, such as pension funds, university endowments, and sovereign wealth funds. A single controversy can quickly tarnish that reputation, particularly when investors are found to own companies that harm society or the environment.

The Sierra Club's *Investing in Carbon Bombs* report brought this risk into sharp focus [150]. It revealed how five major asset managers – BlackRock, Vanguard, JPMorgan Asset Management, State Street, and Invesco – used people's retirement savings to finance fossil fuel expansion projects. Among these was the Willow Project, a vast oil drilling initiative in Alaska's National Petroleum Reserve, which environmentalists condemned as '*climate sabotage*'. The viral #StopWillow campaign amplified public outrage. Despite their climate pledges, these investment firms collectively allocated USD 1.6 billion to new bonds for oil and gas companies in just 17 months. The report accused them of enabling '*the very activities that exacerbate risk for their clients' portfolios*', warning that such investments could become stranded assets. The backlash was swift, with growing calls for asset managers to stop financing fossil fuel projects and fulfil their fiduciary duties by aligning with climate goals.

Other campaigns have taken aim directly at asset owners. In 2025, more than 200 Hollywood actors – including Jane Fonda, Mark Ruffalo, Don Cheadle, and Rosario Dawson – launched the Retire Big Oil campaign [151]. Their demand: that the SAG-Producers Pension Plan divest its USD 100 million in fossil fuel holdings. Managing USD 5 billion for 65,000 union members, the fund was accused of supporting industries driving climate collapse. The actors did not mince words, calling the investments '*morally indefensible*' and '*like betting on a Blockbuster in a Netflix world*'. Their campaign urged trustees to divest from oil, gas, and coal, reinvest in climate-safe funds, and improve transparency and member engagement.

To avoid such reputational risks, some investors have made sustainability central to their strategy. Dutch pension fund PGGM, for example, has embedded the principle 'know what you own' into the heart of its investment policy. As Piet Klop, PGGM's Head of Responsible Investment, explained, the fund aims to '*reacquire the ability to explain*' the purpose of each investee company – based not only on financial fundamentals but also on sustainability performance [152].

Sustainable investing helps investors understand the social and environmental footprint of the companies they own. Every company in the portfolio has been screened. Those that make the cut are there for a reason: either they meet minimum sustainability standards, or their financial performance is compelling enough to justify inclusion despite adverse impacts.

But sustainable investing offers more than just transparency and less existing negative impacts in the portfolio. It also reduces the likelihood of owning companies that become embroiled in *future* controversies [153].

My own research with Anna Vasileva and Laurens Swinkels shows that firms that are better aligned with the Sustainable Development Goals (SDGs) are less likely to be involved in scandals – whether related to corruption, fraud, or environmental violations [153]. Over time, sustainable companies tend to face fewer controversies, and when issues do arise, they are typically less severe and touch on fewer sensitive areas. Sustainable investing thereby helps not only to 'know what you own', but it also 'anticipates what you own'.

In short, investing in more sustainable companies lowers the risk of unpleasant surprises. From a reputation management perspective, this matters. It is costly – both financially and socially – for investors to explain to the public, and to the ultimate owners of capital, why they hold shares in companies that cause harm. Just as sustainable investing presents a 'fear of missing out' on strong returns of companies you may not have invested in, it also presents a 'fear of being in' – of being exposed to companies likely to become scandalous.

7.4 Looking into the Past or Anticipating the Future?

So far, this chapter has mainly looked backwards, examining whether sustainable investing has helped or hurt financial performance in the past. Backtests, with decades of data, offer the most robust way to assess outcomes. But in the context of sustainability, does relying on history truly make sense?

Not really. Looking backwards assumes the future will mirror the past. Sustainability, by contrast, is dynamic, complex, and constantly evolving.

If sustainability matters today, its relevance will only grow as the world changes. Physical risks from climate and nature loss are intensifying. Policymakers are rewriting the rules of the game. Courts are holding companies – and the investors who finance them – accountable for their impacts. Together, these forces are reshaping the investment landscape. In this new reality, sustainability is not optional; it is essential for navigating the future.

7.4.1 Physical Risks

Scientists warn that humanity is breaching multiple planetary boundaries – thresholds that define the safe operating space for life on Earth.[1] Crossing these boundaries does not just degrade ecosystems; it destabilizes the systems that underpin economic activity, social wellbeing, and financial markets. As the *Planetary Health Check 2025* report explains, we are entering a new reality – marked by more frequent climate disasters, collapsing biodiversity, and rising health risks [154]. These changes are neither linear nor predictable. They trigger feedback loops and tipping points that can lead to abrupt and irreversible shifts in the Earth system.

[1] The nine planetary boundaries are: (1) Change in biosphere integrity; (2) Introduction of novel entities; (3) Modification of biogeochemical flows; (4) Climate change; (5) Land system change; (6) Freshwater change; (7) Ocean acidification; (8) Increase in atmospheric aerosol loading; and (9) Stratospheric ozone depletion. Only the last two are still in the safe operating zone. Numbers one to three are in a high-risk zone. Numbers four to seven are in an increasing risk zone. Scientists concluded that the first seven of these planetary boundaries have been breached.

For investors, this means the assumptions behind traditional portfolio construction – stable climate, predictable resource flows, insurable risks, unpriced externalities – are no longer valid. Breaching planetary boundaries threatens the resilience of supply chains, the viability of business models, and the insurability of assets. It raises the risk of stranded investments and systemic shocks. In other words, planetary instability translates into financial instability [155].

These risks are not theoretical – they are already disrupting industries. In July 2025, extreme heat across major U.S. airports caused over 7,000 flight delays and hundreds of cancellations, hitting airlines like Delta, American, United, and Southwest [156]. In 2022 and 2023, droughts in South America disrupted soybean and corn production, hurting agricultural companies [157]. Hurricane Otis struck Acapulco in 2023, damaging 80% of hotels and causing USD 16 billion in losses [158]. Shipping companies increasingly face dangerously low water levels on rivers like the Rhine and the Mississippi, driving up costs across industries [159]. Water scarcity threatens Taiwan's semiconductor industry: during the global chip shortage in 2021, a severe drought hampered production at TSMC and other major firms, highlighting the sector's vulnerability to environmental stress [160].

Examples like these will become more frequent. Climate scientists warn that with further warming, every region will face escalating hazards [161]. Tipping points that lead to large and often irreversible shifts amplify this risk [162]. For instance, melting ice sheets reduce sunlight reflection, accelerating warming; while thawing tundra releases methane, a potent greenhouse gas, fuelling a vicious cycle. Investors will increasingly feel these impacts through rising physical risks to portfolios.

The financial implications are stark. A World Economic Forum study estimates that climate hazards could cut corporate earnings by 6.6%–7.3% annually by 2035, causing USD 560–610 billion in losses per year [163]. Nature loss adds further pressure: more than half of global GDP depends on ecosystem services, and food-related companies could lose up to 26% of their value by 2030 due to degraded soils and disrupted water cycles [164]. UNCTAD warns that overfishing, pollution, and biodiversity loss threaten the USD 2.2 trillion ocean economy [165].

In light of these growing physical risks, the insurance sector – traditionally the first line of defence – is retreating. Allianz's Günther Thallinger warns: '*We are fast approaching temperature levels – 1.5°C, 2°C, 3°C – where insurers will no longer be able to offer coverage for many of these risks*'. When insurance dries up across economic sectors, markets falter. Without insurance, assets become unfinanceable. As Thallinger puts it: '*This is not about saving the planet. This is about saving the conditions under which markets, finance and civilization itself can continue to operate*' [166].

Further evidence comes from a 2023 study, *Rising Temperatures, Falling Ratings*, which shows climate change could trigger sovereign credit downgrades as early as 2030 [167]. These downgrades would raise borrowing costs by USD 45–203 billion annually for governments and USD 10–61 billion for companies – depending on the warming scenario. Optimistically, the study finds that stringent climate policy aligned with the Paris Agreement could nearly eliminate these effects.

The takeaway is clear: physical risks from climate change and nature loss are no longer distant – they are here, and they will intensify. For investors, this means sustainability is not a side issue; it is central to protecting portfolios and navigating a rapidly changing world. This is amplified by the responses of policymakers to this changing reality.

7.4.2 Inevitable Policy Responses

How will policymakers react as climate change, biodiversity loss, and socio-economic risks intensify? Their responses will reshape the investment landscape, creating both risks and opportunities that investors cannot ignore.

At the global level, frameworks like the Sustainable Development Goals (SDGs), the Paris Agreement, and the Kunming-Montreal Global Biodiversity Framework (GBF) chart an ambitious course for sustainability. Yet, progress remains far too slow [168–171]. Experts warn that incremental change will not suffice. Bold, coordinated action is needed to remove roadblocks that deteriorate outcomes and accelerate progress [172], by translating these global commitments into national policies, regulations, and laws.

For investors, this means that policy responses will create both transition risks and opportunities. Companies that fail to adapt may face regulatory penalties, stranded assets, or reputational damage. Those aligned with sustainability goals – such as clean energy providers, sustainable agriculture firms, and inclusive tech innovators – stand to benefit from powerful policy tailwinds. Understanding these responses is not just about managing risk; it is about positioning for long-term value creation.

Consider the automotive sector. Governments worldwide have set sunset dates for internal combustion engines (ICE). Norway leads with a full phase-out by 2025. Iceland, Ireland, Israel, the Netherlands, Slovenia, Sweden, and Singapore follow in 2030. By 2035, all new cars must be electric in California,

Canada, Chile, Japan, New South Wales, and the EU. China plans to end ICE sales by 2035, allowing hybrids until 2045.

It is hard to overstate the impact of policies like these. They are transforming the entire automotive industry – from automakers to suppliers and infrastructure providers. This transition is already underway. In Europe, nearly six out of every 10 new cars sold are electric or hybrid [173]. EV and hybrid sales are surging with double-digit growth, while petrol and diesel car sales are in freefall – dropping over 20%. For legacy manufacturers, the transition is costly and complex. Volkswagen [174] and Mercedes-Benz [175] are accelerating, with roughly 10% of sales now electric. Ford [176] lags at 5%. Failure to adapt could mean losing access to key markets and, ultimately, survival. Meanwhile, pure-play EV firms like Tesla and BYD enjoy a structural advantage: they have no legacy ICE business to unwind.

A similar playbook is applied to the energy sector. Recognizing that burning thermal coal is a leading cause of climate change prompted many governments to adopt phase-out dates. Several U.S. states – including Washington and Oregon – require utilities to eliminate coal-fired power generation by the 2030s. Hawaii banned new permits for coal-burning facilities in 2022. In Europe, multiple countries are phasing out coal by the early 2030s. China has committed to peak coal consumption by 2025, replacing it with renewables at scale. These regulations force utilities – and the mining firms that supply them – to transition their business models to survive.

Other sectors with negative externalities are also facing increasing regulation. Packaging companies in the EU must shift to fully recyclable packaging by 2030 [177]. Food companies face policies aimed at improving nutritional value and reducing harmful ingredients [178]. Fashion companies are increasingly banned from using toxic substances like 'forever chemicals' [179]. And construction companies are increasingly held accountable for meeting environmental performance standards [180].

Policymakers are not only phasing out harmful practices. They are also incentivizing sustainable ones. EV subsidies, tax credits, grants, and infrastructure investments are common. These measures accelerate adoption and reward companies already aligned with the transition.

The implications for investment portfolios are clear. Policies create winners and losers. They also create stranded assets: investments that lose value as regulations tighten [181]. A car factory producing ICE vehicles or a coal-fired power plant may become a liability overnight. The rules of the game are changing, and with them, the risk of obsolete business models.

And policy is only part of the story. Clear rules also make companies legally accountable, creating litigation risks.

7.4.3 Legal Action

In May 2021, a Dutch courtroom made history. Milieudefensie, a climate advocacy group, sued Shell – not for an oil spill or a broken contract, but for its business model. The claim was bold: Shell's greenhouse gas emissions and lack of meaningful climate action violated its duty of care under Dutch civil law and threatened human rights. Drawing on climate science and international agreements, the plaintiffs argued Shell had an '*unwritten duty of care*'.

The Hague District Court agreed. In a landmark ruling, it ordered Shell to cut global CO_2 emissions by 45% by 2030, relative to 2019 levels – and included Scope 3 emissions, those from burning Shell's products. For the first time, a private company was legally required to align its strategy with the Paris Agreement. Shell appealed, and in 2024 the Court of Appeal upheld the duty of care but removed the enforceable target. Even so, the case remains a milestone: climate litigation has moved from governments to corporations.

Financial markets are paying attention. A study of 108 climate lawsuits (2005–2021) shows that litigation risk is now priced in [182]. On average, a company's stock price falls by 0.41% following a climate-related filing or an unfavourable court decision. The largest fossil fuel producers are hit the hardest: their stock drops by 0.57% following filings and 1.50% after unsupportive court rulings.

Science is amplifying this trend. Researchers have developed methods to link emissions from individual companies to specific climate damages, which is a breakthrough for liability claims. Consider the heatwaves that hit Europe in summer 2025, which caused massive wildfires in Spain, infrastructure to crack in Germany, and energy reactors to shut down in Switzerland – among other adverse impacts. How can you say what part of these damages has been caused by the emissions of one firm, and not by another company?

In a recent Nature study, scientists proclaim that '*drawing quantitative linkages between individual emitters and particularized harms is now feasible, making science no longer an obstacle to the justiciability of climate liability claims*'. Their punchy conclusion is that '*the scientific case for climate liability is closed*' [183].

Their findings are staggering. Extreme heat linked to emissions from 111 companies cost the global economy USD 28 trillion between 1991 and 2020.[2]

[2] In their paper, they combine the greenhouse gas emissions of major fossil fuel companies with the latest climate modelling. This way they are able to compare what Earth's climate would be like without the emissions caused by a company with the current state of affairs. The difference between these provides a causal link. This is known as a 'but for' standard – which allowed the scientists to say that a particular part of a climate-related hazard would not have occurred 'but for' the emissions generated by this or that individual firm.

Just five firms account for USD 9 trillion of those losses. Chevron tops the list, responsible for an estimated USD 791 billion to USD 3.6 trillion in heat-related damages. Compare that to Chevron's 2024 earnings: USD 45 billion.[3] If courts impose compensation based on these figures, the company – and its investors – face existential risk.

To date, polluters have not paid for these damages. But with mounting legal action and scientific attribution, that could change. And climate is only the beginning. Similar liability frameworks are emerging for other environmental harms, such as pollution.

A *Science* study analyzed data from 84 countries over five years to identify the companies most responsible for branded plastic waste [184]. While half of the items lacked identifiable branding, researchers found that just five firms – Coca-Cola, PepsiCo, Nestlé, Danone, and Altria – account for a quarter of all branded pollution. The authors concluded that tackling global plastic pollution will require policies to phase out nonessential single-use products, directly impacting the business models of these companies. As with climate liability, this research lays the groundwork for estimating financial damages and holding firms accountable through regulation and litigation.

The same pattern applies to chemical pollution. Legal action over 'forever chemicals' is surging. These substances – known as PFAS – do not break down in nature or the human body, contaminating water, soil, and food while posing severe health risks [185]. PFAS have been widely used in non-stick cookware, rainwear, and firefighting foam.

The financial stakes are enormous. In August 2025, DuPont agreed to a USD 2.5 billion settlement with the state of New Jersey over decades of PFAS contamination [186]. Other chemical giants – including 3M, BASF, Tyco, and Chemguard – face similar lawsuits, with settlements reaching into the billions. A federal judge overseeing PFAS litigation described the exposure as 'an existential threat' to some companies [187]. European courts are following suit, with cases filed in Belgium, France, Germany, Italy, the Netherlands, and Sweden [188]. Even the Dutch fishing industry has sued chemical firms for tens of millions in damages after PFAS contamination rendered fish unmarketable.

These risks might seem remote from day-to-day investment decisions. But a survey of 439 institutional investors suggests otherwise [189]. Researchers found that institutional investors believe climate risks carry significant financial

[3] In 2024, Chevron's earnings before interest, taxes, depreciation, and amortization (EBITDA) equalled USD 45.3 billion, according to the investments website macrotrends (https://www.macrotrends.net/stocks/charts/CVX/chevron/ebitda)

implications for portfolio firms, driven primarily by regulatory change. Half of the respondents said that regulatory shifts due to climate change already pose relevant risks today – not at some distant point in the future.

7.4.4 Looking Ahead

The evidence reviewed in this chapter looks backward – at what sustainable portfolios have delivered historically. But history is an unreliable guide to a future shaped by escalating physical risks, tightening regulation, and growing legal accountability of the kind described above.

Investing sustainably means building portfolios that anticipate this changing world, shaped by escalating physical risks from climate change, biodiversity loss, and inequality, as well as evolving policy responses that reshape market dynamics. And it is not just about future-proofing financial performance; it is also about actively shaping real-world outcomes and contributing to long-term change.

In this light, the future performance of sustainable investing may diverge significantly from its past. The age-old investment disclaimer – '*past performance is not indicative of future results*' – is especially relevant here.

7.5 Concluding Remarks: A Tale of Two Investors

Sustainable investing will only gain traction if it delivers financial performance. The good news? So far, there is no need to worry.

Evidence shows that sustainable portfolios tend to perform on par with conventional ones. They offer sufficient diversification and maintain exposure to well-established drivers of return – value, momentum, and quality. These results hold over the long term, provided the investment universe is not narrowed too aggressively. Looking back, sustainability has not consistently boosted returns, but it has not come at a cost either.

The real upside lies elsewhere. Sustainable investing offers real and relevant benefits. This includes emotional gains associated with knowing that money is invested in line with clients' moral values. There are also expressive benefits, driven by more responsible portfolios that have a lower risk of financing scandals, thereby harnessing investors' reputation.

And what matters too is that sustainable investing prepares portfolios for a changing world. As we continue to breach planetary boundaries and face rising political volatility, back-tests are no longer reliable predictors of future performance. Sustainable investing equips portfolios for a world that is no longer stable – and helps shape a future that can be. Sustainable investors look forward rather than dwell on the past.

In this changing world, investors are at a crossroads. The famous opening to Charles Dickens' *A Tale of Two Cities* expresses the two avenues that are available:

> '*It was the best of times, it was the worst of times, it was the age of wisdom, it was the age of foolishness, it was the epoch of belief, it was the epoch of incredulity, it was the season of Light, it was the season of Darkness, it was the spring of hope, it was the winter of despair, we had everything before us, we had nothing before us…*'

One road that investors can take is marked by incredulity. Investors can ignore how sustainability is affecting their investments and how the world around them is changing. The other road is characterized by wisdom: position their portfolios to avoid sustainability risks, seize opportunities, and be resilient to rising physical risks and growing policy responses. The former road likely leads to a season of Darkness, the other to Light.

8

The Sustainability of Nations

Can Investors Help Countries Finance Sustainable Development?

Thus far, we have focused on investing in companies. Yet more than a quarter of all invested capital is allocated to bonds issued by governments. This makes 'sovereign' bonds the second-largest asset category of the past 50 years [190]. Given this significance, this chapter shifts the focus from corporate investments to sovereign ones – exploring how governments finance sustainable development and how investors can help close the financing gap that continues to stall progress.

From the outset, the challenge is clear: advancing sustainable development requires massive investment. While the SDGs are considered attainable at a reasonable cost, these global goals for sustainable development face an annual USD 2.6 trillion financing gap [191] – roughly 2.7% of global GDP, equivalent to the economy of Italy.

This gap is not caused by a lack of capital – there is plenty. The problem lies in the cost of accessing it [192]. Countries most in need of financing simply cannot afford it.

Consequently, the financing gap for the SDGs is particularly wide in poorer nations. For low-income countries such as Vietnam, Rwanda, and Benin, the International Monetary Fund (IMF) estimates that closing the gap requires additional spending equivalent to 15.4% of GDP every year – a figure that dwarfs what most of these countries can realistically mobilize [191]. Emerging market economies like Guatemala and Indonesia face a somewhat lower but still substantial burden of 4% of GDP annually. In contrast, high-income countries face much lower additional spending needs, typically below 1% of GDP. Their challenge is not a monetary one, but one of ensuring that policies translate into meaningful outcomes.

J. A. van Zanten, *Big Money, Better World?*,
https://doi.org/10.1007/978-3-032-32442-9_8

Table 8.1 Development indicators for top ESG-rated countries[a]

Country	Life Expectancy at Birth (Years)	Expected Years of Schooling (Years)	Maternal Mortality Ratio (Per 100,000 Live Births)	Infant Mortality Rate (Per 1,000 Live Births)	Access to Electricity (% of Population)	Access to Sanitation (% of Population)
Denmark	82	19	4	3	100	99
Finland	82	19	8	2	100	90
Norway	83	19	1	2	100	78
Sweden	83	19	4	2	100	95

[a]Statistics in the tables presented in this chapter are collected through the excellent World Bank's World Development Indicators database, which can be accessed at https://data.worldbank.org/

Across these indicators, a consistent picture emerges: top-rated countries have not merely achieved strong outcomes in one or two areas, but across all dimensions simultaneously – longevity, education, health, and access to basic services. Sustainable success, it seems, requires getting many things right at once.

A child born in these countries today can expect to live roughly 82 years. Scandinavians typically spend about a quarter of their lives in education. Access to healthcare is universal, contributing to some of the world's best health outcomes: maternal and infant mortality rates are exceptionally low, for instance.

Nearly all inhabitants have access to electricity, with renewables accounting for 53% of energy output in Finland and up to 99% in Norway. Most people also enjoy access to safely managed sanitation services, though rural communities in Norway and Finland face some persistent challenges.

According to the UN Development Programme, these development gains are broadly shared across genders, resulting in comparable levels of well-being for men and women.

Finally, the Scandinavian countries are characterized by robust democratic institutions, political stability, and low levels of corruption, reinforcing their strong performance in sovereign ESG assessments.

8.1.2 Mid ESG-Rated Countries

The middle of sovereign ESG rankings is both large and highly diverse. It includes major emerging markets such as Brazil, India, China, and South Africa, each with substantial populations and complex development trajectories. Alongside these giants are many smaller states. Jordan, Bahrain, and Kuwait represent Middle Eastern economies in this group. Several Asian nations – Malaysia, Thailand, Vietnam, and Indonesia – also appear in this mid-tier. And a range of African countries, including Ghana, Senegal, and Rwanda, have made notable progress in recent years, lifting them into the middle of the distribution.

This diversity is reflected in their performance across key development indicators (Table 8.2).

What unites this diverse group is not a shared level of development, but a shared position in transition. These are countries that have made meaningful progress in some areas, while facing persistent gaps in others. Their middle ranking reflects not mediocrity but complexity.

Life expectancy varies widely across the middle tier of sovereign ESG performers. A person born in Ghana today can expect a noticeably shorter life than someone born in Jordan, Malaysia, or China, where life expectancy approaches levels seen in high-income countries. In Kuwait, life spans are even comparable to those in Scandinavia.

Educational opportunities follow a similar pattern. Children in countries like Vietnam and China typically spend well over a decade in school, while those in Indonesia, Jordan, Rwanda, and Ghana receive fewer years of formal education.

Table 8.2 Development indicators for mid ESG-rated countries

Country	Life Expectancy at birth (years)	Expected years of schooling (years)	Maternal mortality ratio (per 100,000 live births)	Infant mortality rate (per 1,000 live births)	Access to electricity (% of population)	Access to sanitation (% of population)
Bahrain	81	16	17	7	100	93
Brazil	76	16	67	13	100	55
China	78	15	16	5	100	69
Ghana	65	11	234	28	90	19
India	72	13	80	25	100	63
Indonesia	71	13	140	17	99	-
Jordan	78	13	31	12	100	80
Kuwait	83	16	8	8	100	100
Malaysia	77	13	26	7	100	86
Rwanda	68	13	229	31	64	-
Senegal	69	9	237	30	74	30
South Africa	66	14	118	24	88	74
Thailand	76	15	34	8	100	27
Vietnam	75	15	48	14	100	45

Health outcomes likewise range sharply. Some middle-tier countries have relatively low infant and maternal mortality rates, while others continue to struggle with preventable deaths, reflecting major differences in healthcare access and quality.

Access to basic services also diverges. Electricity, clean water, and sanitation are nearly universal in places such as Kuwait and Malaysia. Coverage is substantial, though not complete, in China, Jordan, and South Africa, and continues to expand in Indonesia, Rwanda, Ghana, and Senegal.

Governance further shapes these development paths. Some countries are led by strong, autocratic rulers, while others operate more democratic systems. And although corruption is a challenge in several of them, others have made progress in strengthening public institutions and improving accountability.

8.1.3 Low ESG-Rated Countries

At the bottom of ESG ratings for countries, you will find countries facing persistent development challenges.

This includes countries with a recent or continuing history of violent conflict, like Yemen, Sudan, Iraq, Mali, Burkina Faso, or the Democratic Republic of Congo. It also includes countries that have faced persistent development challenges, like Madagascar or Honduras. Russia is ranked poorly by most providers, especially since its war efforts in Ukraine.

Good governance is a common challenge in most of these countries. Some have authoritarian regimes that offer limited freedom to their citizens – like Russia. Others, like Mali, are under military rule, with political parties being banned, dissent being silenced, and elections delayed. Power struggles are commonplace, often harming not only political processes but also people's lives and livelihoods.

Indeed, in the countries ranked at the bottom of ESG ratings, life is much more of a struggle compared to better-ranked nations. Table 8.3 shows the same statistics for these countries.

The indicators show that these countries share a story of fragility: basic systems, including health, education and energy, remain unreliable and out of reach for many. That fragility translates directly into investment risk, which is precisely why they cluster at the bottom of sovereign ESG ratings.

For many people in this group of countries, life is short. Average life expectancy often hovers just above 60 years, with only a few exceptions – such as Honduras, Russia, and Iraq – where people can expect to live noticeably longer. Schooling also occupies only a small share of life. Children in Mali,

Table 8.3 Development indicators for low ESG-rated countries

Country	Life Expectancy at Birth (Years)	Expected Years of Schooling (Years)	Maternal Mortality Ratio (Per 100,000 Live Births)	Infant Mortality Rate (Per 1,000 Live Births)	Access to Electricity (% of Population)	Access to Sanitation (% of Population)
Burkina Faso	61	9	242	45	22	25
Congo, Dem. Rep.	62	11	427	45	22	13
Honduras	73	10	47	13	96	-
Iraq	72	12	66	21	100	51
Madagascar	64	9	445	44	39	13
Mali	60	7	367	58	55	20
Russia	73	13	9	4	100	58
Sudan	66	9	256	39	66	-
Yemen	69	7	118	35	84	43

Yemen, and Burkina Faso typically spend less than a decade in the classroom, while those in Iraq enjoy significantly more years of education than most of their peers in this group.

Healthcare challenges are widespread and, in some places, severe. Maternal and infant mortality rates remain alarmingly high in countries such as Mali, Madagascar, Burkina Faso, and Yemen. These outcomes reflect persistent gaps in prenatal care, limited access to medical facilities, and the overall fragility of health systems.

Access to essential services shows an equally stark divide. In several countries, only a small fraction of the population has access to electricity or safely managed sanitation. Others fare somewhat better, but even the strongest performers in this group fall short of universal coverage. Basic needs, in short, are far from guaranteed for most people living in these countries.

8.1.4 Income Bias

These country examples reflect the Anna Karenina principle, drawn from the famous opening of Tolstoy's novel: '*All happy families are alike; each unhappy family is unhappy in its own way*'. This principle colourfully illustrates that success requires the simultaneous satisfaction of multiple conditions, while failure can result from the breakdown of any single one.

In the context of sovereign ESG performance, this maps directly onto what the data shows. Top-ranked countries like those in Scandinavia tend to share a common set of strengths – strong institutions, universal access to basic services, and equitable development. To score highly, a country must perform well across all of these dimensions at once. In contrast, countries at the bottom of the rankings struggle for diverse and complex reasons: conflict, poor governance, or lack of access to healthcare, education, or energy. Any one of these failures is sufficient to drag a country's overall rating down, regardless of how it performs elsewhere.

This indicates that there are limited paths to success. For a country to be regarded as a top performer, it must meet many positive conditions simultaneously. But there are many paths to failure. Given the many variables scrutinized by sovereign ESG ratings, failure in one area will already drag a country's score downward.

Does this mean that sustainable investors should simply allocate money to the better-ranked nations – like Scandinavia – and avoid emerging markets or struggling states altogether?

The answer is more nuanced and requires investors to confront the ingrained income bias embedded in sovereign ESG ratings. The World Bank thoroughly investigated sovereign ESG ratings. Their research reveals that better-scoring countries tend to share one common characteristic that explains much of their strong performance across sustainability dimensions: money.

In fact, the World Bank finds that about 90% of a country's ESG rating can be explained by its national income, based on regression analysis across major ESG rating providers [198]. Wealthier countries typically have stronger institutions, greater prosperity, better infrastructure, and more equality. As a result, the World Bank argues that sovereign ESG ratings suffer from an '*ingrained income bias*', raising doubts about whether these scores truly reflect a country's sustainability performance.

This income bias poses a serious challenge for sustainable investors targeting government bonds. Sovereign ESG ratings can mask the sustainability risks and opportunities of prosperous countries, while exaggerating those of developing nations. Moreover, relying on these ratings can create, in the World Bank's words, '*potentially perverse investment incentives that drive capital away from lower-income toward higher-income countries*'. In other words, investment portfolios built on these ratings will end up favouring richer countries and overlooking poorer ones.

This hedgehog view is deeply problematic, as the World Bank explains, because it is developing countries that most urgently need funding to pursue more sustainable development pathways. Sustainable investors ought to prioritize, not withhold, investments in these countries. The rationale for this is clearly illustrated by the financing gap facing the SDGs.

But as we have seen, sovereign ESG ratings are not designed to guide investors towards countries that are likely to support sustainable development. Their ingrained income bias instead leads investors to target rich countries that are not short on capital, while not saying much about the quality of their policies for promoting further development. Consequently, steering on ESG ratings misdirects capital allocation from a sustainability point of view.

The consequences are concrete and worrying. Take a country like Ghana. As we will see later, this developing country has made significant strides in areas like healthcare, education, and access to basic needs. Its sovereign ESG rating, however, does not reward this progress. What it will reflect, overwhelmingly, is that the country remains poor. A hedgehog investor following that signal withholds capital from exactly the kind of progress the world needs more of. The fox would look at the country with an entirely different lens: not where the country stands today, but how its policies may promote further progress.

In short, the way investors commonly screen countries on their sustainability performance is dominated by a single view: Hedgehogs focus on money, not on a plurality of sustainability issues. So how can investors identify which countries truly deserve prioritization in a sustainable government bond portfolio – beyond what ESG ratings suggest?

8.2 The Fox's Approach to Sustainable Government Bond Investing

I have been making the case that sustainable investors should invest in bonds issued by governments that have good policies for advancing sustainable development. From this perspective, a country's current development status matters less than its potential for progress. What matters for investors is not where a country stands on the development ladder, but how able it is to rise on it. A sustainable investor should not just finance the status quo, but seek to invest in improvements. This is the fox's approach applied to sovereign investing: rather than asking where a country stands, it asks where a country is going – and whether its policies are capable of getting it there.

The fact that Scandinavians have a higher life expectancy, better health outcomes, and stronger education systems than Ghanaians or Jordanians – which are typically reflected in sovereign ESG ratings – is less relevant in this view. What an investor would want to know is whether countries have policies that lead to more progress. If the government of Ghana or Jordan develops great health or education policies that will improve people's quality of life, then that provides a strong sustainability rationale for extending financing to these countries.

But assessing which countries have the right policies in place to drive such progress has proven a tough hurdle to tackle. The idea of investing in progress rather than the status quo can only be implemented when you can identify where that progress is most likely to occur. This is anything but straightforward. Tracking development indicators alone has been described as an '*unprecedented statistical challenge*' [199], and measuring the policies needed to drive transformative change reveals a '*massive data and knowledge gap*' [200].

The adoption of the SDGs in 2015 spurred a range of initiatives aimed at monitoring how well countries are progressing. The High-level Political Forum on Sustainable Development (HLPF), operating within the UN system, oversees global efforts to review and assess SDG advancement. It conducts thematic reviews of specific SDG areas and facilitates Voluntary National Reviews (VNRs), which are submitted by individual member states. These VNRs include both progress updates and descriptions of national policies intended to advance the SDGs.

Despite these efforts, there is still no consistent or reliable way to evaluate and compare the effectiveness of governmental sustainability policies across countries. This lack of standardization makes it difficult for investors to identify which governments are truly committed to sustainable development.

Compounding the issue is a growing disconnect between rhetoric and action: while many governments have quickly adopted the language of the SDGs, evidence suggests that this has not translated into meaningful policy change. A comprehensive review of over 3,000 studies found that the SDGs have had only limited impact on actual government policies, even as they have become embedded in political discourse [170].

Faced with the complexity of evaluating government policies for sustainable development, it is easy to feel overwhelmed. But sustainable investing is about action – not paralysis. While there may be no universally accepted method for assessing policy effectiveness, we can still draw meaningful insights by analyzing the wealth of data that has emerged since the adoption of the SDGs in 2015. There has been a surge in publicly available data, often reported in comparable formats across nations. These indicators offer a valuable – if imperfect – window into the effectiveness of government policies.

Some of these indicators that are available *directly* reflect policy choices. For example, governments report how much of their GDP is allocated to healthcare or education. Data is available on fossil fuel subsidies and on whether countries have ratified key international treaties – such as the Paris Agreement on climate change or conventions related to human rights, pollution, and disarmament. These are clear – direct – signals of a government's policy orientation.

Other readily available indicators can serve as *indirect* proxies for policy effectiveness. Take maternal mortality: if the ratio is declining over time, it suggests that health policies are improving outcomes. If it is rising, it may point to policy failure. These kinds of outcome-based metrics can help investors assess whether governments are not just making commitments, but delivering results.

By looking at these directly and indirectly relevant indicators, a wholly new and foxy way of evaluating countries' sustainable development efforts emerges, which fully contrasts with sovereign ESG ratings. Let us dive into the domains of health and ocean sustainability to see how policies can be evaluated in practice.

8.2.1 Health and Well-being

Maternal and infant mortality offer powerful signals of a country's healthcare effectiveness and, by extension, its commitment to sustainable development. Take Ghana. In 1985, maternal mortality stood at 943 deaths per 100,000 births; by 2023, it had fallen to 234 – a 75% reduction, including a 25% drop since the SDGs were adopted in 2015. Infant mortality shows a similar trajectory: from 122 deaths per 1,000 births in 1960 to 28 today. These gains suggest Ghana's health policies are working. Senegal, Lao PDR, and Chad show similar progress.

But not all countries follow this path. In Madagascar, infant mortality has stagnated at 44 deaths per 1,000 births for a decade. Jamaica's maternal mortality has worsened, rising from 100 to 130 since 2015. These contrasts underscore a critical point: progress matters. Declining mortality signals effective policies. Stagnation or regression suggests the opposite.

High-income countries present a different picture. Sweden's maternal mortality is just four per 100,000 births, and infant mortality two per 1,000 – with the ratios being stable over the past years. The U.S., however, still lags, with rates three to four times higher than Sweden's and, in particular, its maternal mortality rate, showing a worrying spike in 2021–22.

Of course, health is broader than these two indicators. Spending on healthcare, hospital bed availability, and preventable disease mortality all offer valuable insights. For investors, these signals matter. Financing countries where health outcomes are improving – through sovereign bonds – can help sustain progress. In contrast, investing in nations with stagnant or worsening indicators offers less potential for impact.

Health is one dimension. But the fox's approach demands breadth – and the same logic applies across environmental goals, where the data tells an equally instructive story. Let us dive into ocean sustainability as another example.

8.2.2 Ocean Sustainability

I recently enjoyed watching David Attenborough's documentary *Ocean.* It reminds us that the sea is Earth's life-support system, regulating climate, producing half our oxygen, and sustaining biodiversity. Yet it faces mounting threats: pollution, overfishing, acidification, and climate change. Governments pledge action, but how can investors tell if those promises translate into progress?

We can apply a similar approach to ocean sustainability as we did with health. Two indicators stand out as an illustration. The first concerns Marine Protected Areas (MPAs). These are direct policy tools for conserving ecosystems. This is widely recognized as a key strategy for restoring marine populations, habitats, and ecosystems [201–204].

Looking at the most recent figures provides insight into current policy actions. Here, the data shows stark contrasts: France protects nearly 50% of its waters, Ireland just 2% [205]. Yet trends matter too. The U.S. protects 19% today – down from 41% in 2016 – while Belize doubled its coverage to over 20% since 2015. Hence, U.S. ocean conservation policies appear to be weakening, while Belize is strengthening its efforts. And although MPAs are not perfect, they remain a critical tool for promoting ocean sustainability.

The second indicator is bottom trawling: industrial fishing that drags heavy nets across the seabed, destroying habitats and releasing carbon at levels comparable to aviation. The practice, likened to 'bulldozing a rainforest', was captured in *Ocean* in scenes described as '*fit for a horror film*': heavy nets dragging across the seabed, schools of fish scattering, and barren aftermath left in their wake [206]. Despite calls from scientists to ban trawling in MPAs [207], it persists – even in protected zones. This is notwithstanding the strong ecological and economic arguments against the practice: reducing industrial fishing practices allows fish stocks to recover [208] and banning trawling could deliver €8.4 billion in net benefits over two decades [209].

So how can investors understand which countries have more sustainable fishing practices?

Again, data can help. The Sea Around Us (SAU)[3] provides data on fisheries and their ecological impacts [210]. One of its indicators tracks the

[3]This is a research initiative at the University of British Columbia and the University of Western Australia.

proportion of fish caught by bottom trawling and dredging. Its data shows that between 2015 and 2019, 87% of Romania's catch came from trawling – suggesting that a large share of its fishing fleet applies destructive practices.

But is this the full picture?

The scale of countries' fishing sectors also has to be accounted for. World Bank data shows that Romania caught 6,256 metric tons of fish in 2022 – a relatively small amount. In contrast, China landed over 13 million metric tons – over two thousand times more than Romania – with nearly half (48%) caught by trawling. While Romania's fishing sector is less sustainable (a higher share of fish is landed by trawling), China's sheer volume of catch results in far greater ecological impact.

For investors, these indicators – MPA coverage and trawling intensity – offer a window into policy effectiveness. Countries expanding protection and curbing destructive fishing signal stronger sustainability commitments. Allocating capital toward such nations can help finance progress, while avoiding those that undermine ocean health.

8.2.3 Rating Countries

These examples illustrate that existing data can already help investors identify which countries have supportive policies for sustainable development. But what happens when you combine multiple indicators – spanning health, ecosystems, and other sustainability topics aligned with the SDGs – and analyze a broad set of countries?

Together with my former colleague Maria Putintseva, a data scientist with a PhD from the University of Zurich, we analyzed 84 indicators across 170 countries. For each indicator, we assessed whether it directly or indirectly reflects policy effectiveness for sustainable development. We also determined how each indicator could best be interpreted to gain a policy perspective – for instance, by looking at the most recent value or by analyzing trends over time, as we did earlier with mortality rates and marine protection. We then used a machine learning algorithm to synthesize this information and generate a policy effectiveness score for each country. In essence, the algorithm groups countries by their similarity across sustainability indicators – identifying clusters of nations that share comparable policy environments – and scores each group based on how supportive those environments are of sustainable development. This allowed us to move beyond subjective judgments, letting the patterns in the data determine which countries are driving progress and which are not.[4]

The results struck us. Only one in five countries have policies that are, on balance, supportive of sustainable development. About 30% maintain policies that are largely neutral – neither clearly advancing nor obstructing progress – while roughly half of all countries risk undermining the SDGs through their current policy choices. Put differently: the policies of roughly 130 countries – home to the vast majority of the world's population – are either neutral or actively working against the goals the global community committed to in 2015. Fig. 8.1 shows how these scores are distributed globally.

The breakdown by individual SDG makes this picture more nuanced – and in some ways more worrying. Policies targeting human well-being show the most positive results. Between 39 and 56% of countries support progress on goals such as poverty eradication (SDG 1), health (SDG 3), education (SDG 4), and gender equality (SDG 5). In contrast, environmental and governance goals face significant challenges. For example, 71% of countries exhibit policies that hinder progress on Life on Land (SDG 15), 53% on Peace, Justice, and Strong Institutions (SDG 16), and 52% on Responsible Consumption and Production (SDG 12).

We compared our scores to sovereign ESG ratings and national income levels. The correlations were only moderate.[5] While countries at the bottom of ESG rankings – and those with very low-income levels – often struggle across the board, many emerging markets with relatively stable governance performed well in several areas. These countries received the highest scores in our assessment. Meanwhile, many high-income countries showed more neutral policy environments. This is partly because they already perform well in many areas, but also because they often fail to drive transformative change in others — particularly in climate and nature.

Of course, ours is just one assessment, and alternative approaches are possible. Still, our findings align with other evaluations – such as those by the UN and the Sustainable Development Solutions Network (SDSN) – which suggest that global progress on sustainable development is stagnating or even reversing. We published our results in an academic journal to contribute to the growing debate on how best to evaluate government sustainability policies [171].

[4]The algorithm we used is called BIRCH (Balanced Iterative Reducing and Clustering using Hierarchies).

[5]To be precise, the correlation between our sustainable development policy rating and the sovereign ESG rating from Robeco was 0.48. Its correlation with GNI per capita was 0.25.

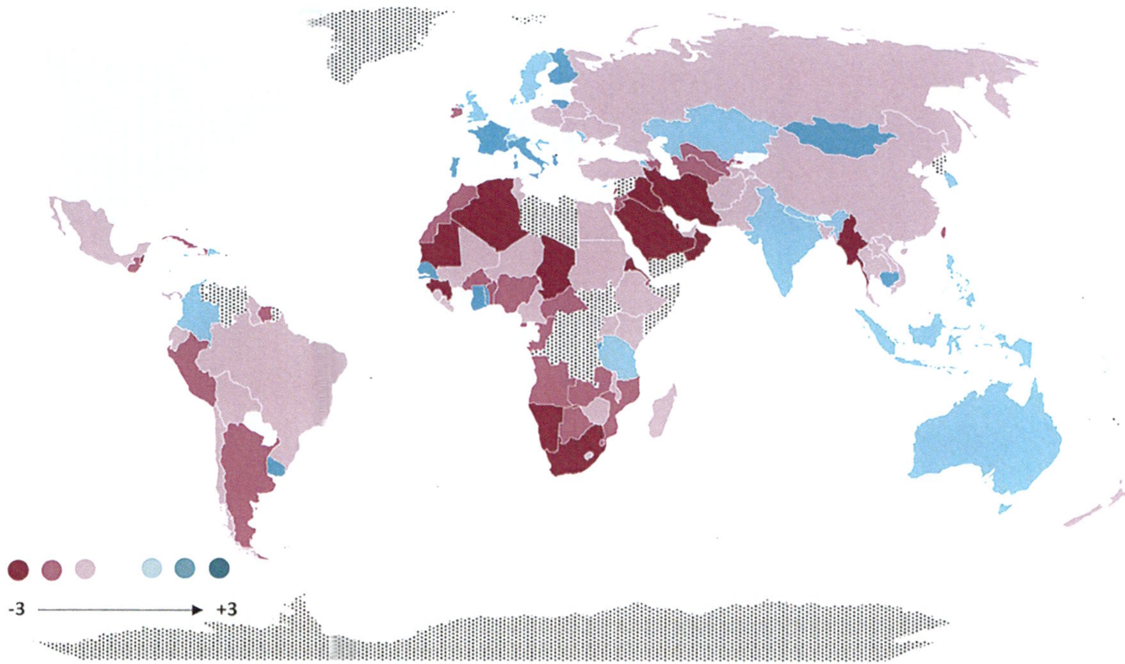

Fig. 8.1 Global distribution of SDG policy scores. This figure shows the distribution of country-level SDG policy scores around the world. The scores range from -3 (low – dark red) to +3 (high; dark blue). Neutral policy scores are shown in light blue. Countries that lack a score – due to insufficiently available data – are shown in black dots. Source: van Zanten, J. A., & Putintseva, M. (2025). Evaluating governmental policies for the sustainable development goals using hierarchical clustering. *International Journal of Sustainable Development & World Ecology*, *32*(3), 322–340.

But this analysis was not just an intellectual exercise. We also explored how these insights could be applied in practice – specifically, to build sustainable government bond portfolios. This led us to conclude that while government bonds can play a key role in financing positive change, they also face important limitations.

Let me show you.

8.3 Investing in Countries with Good Development Policies

There are nearly 200 countries in the world. For 170 of them, we were able to gather sufficient information to assess the effectiveness of their sustainable development policies. But not all of these countries issue bonds that investors can purchase. This makes them effectively non-investable from a government bond perspective.

Leading index providers – such as Bloomberg and J.P. Morgan – help clarify which countries are within reach for government bond investors. A fixed income index is essentially a basket of bonds designed to represent a particular segment of the bond market. These indices are commonly used to benchmark investor performance and to provide insight into market trends.

There are many such indexes. One widely used example is the Bloomberg Global Aggregate Treasury Index, which covers global investment-grade bonds issued in local currency by governments deemed relatively creditworthy. At the end of 2024, this index had a market value of USD 35.6 trillion and included more than 46 developed and emerging countries.

Another index with a different focus is the J.P. Morgan Emerging Markets Local Currency Debt Index. It includes markets with higher risk profiles – such as the risk that a country defaults on its debt obligations, or of currency depreciation – and tracks bonds issued in local currency. Complementing this is the J.P. Morgan hard currency Emerging Markets Bonds Index (EMBI). This index includes 75 emerging market countries. Many of these are unable to raise capital in international markets using their own currency. That is why they issue debt in hard currency – typically the US dollar – which is widely accepted but makes the issuing country vulnerable to exchange rate fluctuations.

Taken together, the countries included by these index providers can be considered investable from a sovereign bond perspective. Those that are not listed are, for practical purposes, out of reach for mainstream investors.

This is where a massive challenge arises for sustainable government bond investors.

These two index providers cover only 112 of the 170 countries we rated. In other words, 58 countries – over one-third – do not have sufficiently accessible government bond markets. This is particularly problematic for countries that are implementing policies likely to lead to positive sustainability outcomes, yet lack access to international capital. These countries are home to 2.4 billion people, underscoring the scale of the problem. Some of these countries are among the most policy-active on sustainable development: various nations in sub-Saharan Africa and South and Southeast Asia have made big strides on health, education, clean energy, and access to basic services, yet remain effectively invisible to mainstream bond investors. Sustainable investors focusing on government bonds cannot support positive policy environments in these regions, nor contribute to improved outcomes through capital allocation. Hence, the financing gap is not just a problem of cost; it is also a problem of reach.

Still, there is reason for measured optimism. Around 2.1 billion people live in countries that have both supportive sustainability policies and are accessible through international government bond markets. So, while the reach of investors is limited, there remains a significant opportunity to finance positive change through government bonds.

Having developed these sustainability policy ratings, I wanted to put them into practice by designing model investment portfolios. To do so, I enlisted the help of Laurens Swinkels, a quantitative investment specialist and Associate Professor at Erasmus University, and we expanded the team with Rikkert Scholten and Bruno Rein. Our goal was to see what a portfolio would look like that invests in countries with supportive sustainable development policies.

We used the earlier discussed Bloomberg and J.P. Morgan investment indices as a starting point to define the investable universe of government bonds. Then, we looked at each country's sustainability policy rating – indicating how supportive its policies are of advancing sustainable development.

To determine which countries to include in our model portfolios, Laurens proposed a straightforward method. We multiplied the market value of each country's bonds by a factor based on its sustainability rating. Ratings were categorized as negative, neutral, or positive – with further distinctions within the negative and positive groups (low, medium, and high).

Countries with negative policy scores were excluded entirely – their bond market value was multiplied by zero. Our belief is that countries whose policies are likely to detract from sustainability goals do not belong in a portfolio aimed at financing progress.

Neutral countries were underweighted, with a factor of 0.5. While these countries may not demonstrate strong sustainability performance in the near future, they are also unlikely to deteriorate – and many may still require access to financing to accelerate development.

For countries with positive ratings, we applied multipliers of 1.0, 1.5, or 2.0, depending on the strength of their policy support. Following this logic, the more supportive a country's sustainability policies, the greater its representation in the portfolio – and thus the more money it is receiving.

To maintain diversification, we capped the overweight for any individual country at 10 percentage points above its benchmark weight. This approach is best understood as a sustainability overlay – a method for tilting a conventional government bond portfolio toward countries with stronger development policies, not a complete investment strategy. Practical considerations that any real portfolio manager would need to address, such as credit risk, currency exposure, liquidity constraints, and duration management, were not incorporated here but can easily be added. In sovereign bond investing in particular, these factors can be as consequential as the sustainability screen itself: a country with excellent development policies may still carry significant default risk or currency volatility that affects the investment case. Our aim was not to resolve these tensions but to demonstrate that a principled sustainability tilt is feasible and that its tracking error is modest. Incorporating these additional considerations would be the natural next step for any practitioner seeking to implement this approach in a live portfolio.

So, when you apply this method, what would your investment portfolio look like?

Let us start with the local currency emerging markets index. Your portfolio would allocate significantly more to smaller markets with strong sustainability policies. For example, 4.75% of your portfolio would consist of bonds issued by Uruguay, even though its index weight is just 0.16%. Similarly, 1.45% would go to the Philippines – whose index weight is only 0.05%.

You would also hold large positions in countries that already make up a substantial part of the index. India and Indonesia, both with supportive sustainability policies, would each account for a maximum 10% of your portfolio.

In contrast, your portfolio would exclude some large countries with negative scores, such as Brazil and China. Brazil scores poorly on protecting land-based ecosystems and corruption, while China faces issues related to human rights and civil liberties. Other excluded countries include Argentina, Egypt, South Africa, and Turkey.

The hard currency emerging markets index contains many smaller countries, given their challenges in issuing debt in local currency. Using this index, over 5% of your portfolio would go to Ghana (0.88% in the index), nearly 2% to Mongolia (0.33%), and over 1.5% to Senegal (0.27%). Other included countries are Sri Lanka, Serbia, and Jamaica. Excluded countries due to weaker sustainability policies include Zambia, the United Arab Emirates, Nigeria, and Kuwait.

If you instead construct a fully global portfolio using the Bloomberg Global Treasury Index, you will overweight countries like Korea, Lithuania, France, and the United Kingdom. Excluded countries would include Israel, Ireland, and Singapore.

These examples demonstrate how a straightforward approach can help investors actively align with sustainable development goals – rather than rely on conventional indices that passively allocate capital to countries undermining sustainability goals.

But can government bonds help close the financing gap for the SDGs, and thereby contribute positively to sustainable development?

8.4 Closing the Sustainable Development Financing Gap

Earlier, we saw that the financing gap for the SDGs is particularly pervasive in lower-income countries, and that it is not caused by a shortage of capital. There is enough money in the world to finance sustainability goals. Instead, the financing gap is a result of the high costs that emerging markets face in accessing capital.

Indeed, the world faces a stark reality: the cost of borrowing for governments in emerging markets is, on average, much higher than for those in developed economies. At the end of 2023, for example, the average interest rate on hard currency debt for emerging markets was 8%, while the US government could borrow at just 4%. This difference is not just a technical detail – it has real-world consequences. It means that a project designed to advance sustainable development might be financially viable in the US, but out of reach in countries that arguably need it most.

This is where the investment approach laid out above can make a real difference. By directing more capital toward government bonds issued by countries with strong, forward-looking sustainability policies – especially in emerging markets – investors can help lower the cost of capital. This, in turn, makes it more affordable for these governments to finance projects that drive progress, from building clean energy infrastructure to improving healthcare and education.

The numbers are clearly positive. At the end of 2023, emerging market governments with at least neutral or positive sustainability policy scores had issued about USD 1 trillion in debt. In contrast, the global investment-grade government bond market – dominated by developed countries – was worth a staggering USD 35 trillion. Even a modest reallocation of investment from developed to emerging markets could dramatically expand the financing available for sustainable development where it is needed most.

Would this help developing countries gain better access to finance? History shows that bond investors have the power to move markets and influence outcomes. When large investors change their allocations, interest rates respond – sometimes quickly and significantly.

A good illustration comes from two researchers, Greenwood and Vayanos, who looked into the United States Treasury buyback program of 2000–2001 [211]. This program took place at a time marked by a few years of budget surpluses. These surpluses led to reduced government financing requirements and thus few new bond listings. At this time, the government sought to gain greater control over when the United States would have to repay its outstanding debt.[6]

To do so, the US Treasury bought back 42 bonds with a combined face value of USD 63.5 billion – some 10% of the December 1999 outstanding value of marketable long-term government bonds, and approximately 2% of all marketable government debt. Greenwood and Vayanos then explained that this program led to '*dramatic effects*': the spread between 20-year and 5-year spot rates fell by 0.65% in less than a month. This is a notable decrease in long-term interest rates and thus a substantial drop in lending costs.

Another, more recent, study found similar results. When Dutch pension funds sold 30-year bonds and bought 20-year bonds – following a regulatory change in the discount rate for pension liabilities – the relative yield was affected by 0.30% [212].

These empirical findings are in line with an essential report by the Columbia Center on Sustainable Investment that investigates the cost of capital for climate and SDG finance [192]. The report argues that if institutional investors reallocate even a small portion of their portfolios toward emerging markets

[6] The U.S. government here aimed at managing what is called the maturity structure of outstanding bonds. The maturity structure defines when the borrower must pay back the money (or, in more technical terms, how long it takes for a bond to repay its principal).

with strong sustainability policies, they can help close a meaningful part of the existing financing gap.

In short, this new approach to government bond investing is not just about building novel portfolios. It is about unlocking capital for countries that are ready to drive sustainable development yet face higher borrowing costs. By aligning investment decisions with sustainability policies, investors can help reduce these costs. While the magnitude of the effect is not groundbreaking, if applied at scale, then the effect is significant – enabling countries to obtain more funding. Investors can thereby play a role in turning the promise of sustainable development into a reality for billions of people.

To see whether the approach is practically feasible, we need to ask one more – by now familiar – question: does this approach come at a cost to investors?

A natural question is whether investing more sustainably means sacrificing financial performance. Because we only have a limited history of the sustainability policy ratings, we cannot yet say anything definitive about historical returns from allocating more capital to governments with strong sustainability policies.

What we can assess, however, is how much a portfolio that excludes governments with negative policy scores would deviate from its benchmark. This deviation is measured by the so-called tracking error, which quantifies how closely a portfolio mirrors its benchmark. A low tracking error suggests performance that closely follows the market, while a higher tracking error indicates greater divergence – something that can be desirable for impact, but is often perceived as riskier by investors.

Over the four-year period for which we have sufficient data, we found that the tracking error for sustainable government bond portfolios were 2.5% and 3% for local and hard currency emerging market portfolios, respectively. For a global portfolio – including both developed and emerging markets – the tracking error was just over 1.1%.

What do these numbers imply?

These tracking errors are low for a global portfolio and modest for emerging markets. What this means for investors is that risk and return objectives should not be seen as a reason to avoid sustainable investing in sovereign bonds. And should an investor want to integrate sustainability in this asset class but stick closer to the benchmark (i.e., reduce the tracking error as much as possible), then there are various other options. For instance, rather than excluding all countries with negative policy scores, investors could choose to underweight them. This would reduce the sustainability performance of the portfolio somewhat relative to not investing in negative-scoring countries, but also lower the relative risk of the portfolio.

There are various ways to allocate more capital to countries with strong sustainability plans – and to those facing limited access to financing – without straying too far from broader market benchmarks.

8.5 Engaging with Countries

Can investors do more than just finance countries' sustainability transitions? Obviously, and fortunately, investors cannot vote in national elections. But can they use their voice to engage with governments – much as they do with companies – and encourage progress on sustainable development? The answer is yes – and it is already happening.

Sovereign engagement is the practice of investors using their influence to encourage governments to adopt more sustainable policies. While this may sound ambitious, it is not entirely new. Investors have long interacted with governments on issues like debt restructuring, market access, and regulatory clarity. What is changing is the purpose: moving from protecting financial returns to promoting sustainability outcomes.

This shift is both logical and necessary. Governments are the primary actors in achieving sustainability objectives like the SDGs. They set environmental regulations, fund public services, and shape the legal and institutional frameworks that underpin sustainable development. If investors want their capital to support progress on the SDGs, engaging with governments is a natural next step.

But what does this look like in practice?

One of the most prominent examples is the Investor Policy Dialogue on Deforestation (IPDD). This initiative brings together over 60 financial institutions, representing more than USD 8 trillion in assets, to engage with governments in countries where deforestation is a major concern – such as Brazil and Indonesia. The goal is not to dictate policy, but to express investor expectations and encourage stronger environmental governance. In Brazil, for instance, the IPDD has engaged with senior government officials to raise concerns about rising deforestation rates in the Amazon. Investors have emphasized that continued environmental degradation could undermine Brazil's long-term economic prospects and its attractiveness as an investment destination. These conversations have helped elevate environmental issues on the political agenda and signal that sustainability matters to global capital markets.

As one investor explained, '*It is important to make it clear to the government that we want to contribute to the country – for us, this means that we face lower risks and that we help clients achieve their sustainability ambitions. For the country, there is the benefit that it can improve their risk profile, and thus reduce their lending fees*' [213]. Olga Hancock, Co-Chair of IPDD Indonesia and the Deputy Head of Responsible Investment at Church Commissioners for England, put it succinctly: '*We are supporting the efforts of the Indonesian government in reducing deforestation, and in doing so signal to governments globally that the international flow of capital supports the sustainable development agenda*' [214].

Following its launch in 2020, the IPDD gained significant traction for its work on deforestation. Yet while Brazil and Indonesia are large markets relative to other emerging economies, the lion's share of global fixed income investment still flows to rich countries. Brazil and Indonesia are relevant targets – they certainly face challenges related to deforestation – but what about engaging with those wealthy countries that attract most investment and are also among the leading contributors to global issues like climate change?

Recognizing this, the Principles for Responsible Investment (PRI) launched its Sovereign Engagement Platform on Climate Change in 2023. This ambitious initiative unites more than 40 institutional investors, representing around USD 17 trillion in assets under management, to engage directly with governments on climate action. What sets this platform apart is its global reach: it does not just focus on emerging markets, but also targets major developed economies like Australia, Canada, and the United Kingdom – countries that are both magnets for global capital and significant contributors to global emissions.

Through coordinated dialogues, investors in the PRI platform urge governments to close the gap between current climate policies and Paris-aligned targets, develop credible net-zero transition plans, and improve transparency around climate risks. The platform's pilot engagement in Australia, for example, has already led to constructive conversations between investors and policymakers, helping to put climate risk higher on the national agenda.

Of course, sovereign engagement is not without its challenges. First, the power dynamics are different. Governments are not accountable to shareholders in the way companies are; they answer to their citizens, not to investors. Engagement must be approached with humility and respect for national sovereignty. Investors are not there to impose their will, but to express how certain policies – or the lack thereof – affect their investment decisions.

Second, engagement takes time. Policy change is often slow, especially in complex political environments. Investors must be patient and persistent, recognizing that progress may come in small steps rather than sweeping reforms.

Third, coordination is key. A single investor may have limited influence, but when institutions act collectively – as they do in the IPDD and PRI platforms – their voice becomes harder to ignore. Collaborative engagement also helps avoid mixed messages and ensures that governments receive consistent signals from the investment community.

Finally, credibility matters. Engagement only works if it is backed by action. If investors say they care about sustainability but continue to allocate capital to countries with harmful policies, their words will ring hollow. That is why engagement should be part of a broader strategy – one that includes portfolio construction, transparency, and accountability. This is the sovereign equivalent of the greenwashing problem that haunts corporate sustainable investing: without the discipline to align capital with stated values, engagement risks becoming more signal than substance.

Despite these hurdles, sovereign engagement holds promise. It allows investors to go beyond asset allocation and become active participants in shaping a more sustainable global economy. It also reflects a deeper understanding of how change happens – not just through markets, but through policy.

And perhaps most importantly, it offers a way to bridge the gap between capital and impact. By engaging with governments, investors can help create the conditions that make sustainable development possible – whether that is protecting forests, expanding access to education, or strengthening democratic institutions.

In the end, sovereign engagement shows governments that investors – their financiers – care, and that they can reward positive change. This can be a powerful tool for sustainable investors.

8.6 Concluding Remarks: A Neglected Frontier

Sustainable investing has paid limited attention to government bond markets, despite this being one of the largest asset classes in the world. Investors have focused heavily on companies, while largely overlooking investments in governments. This chapter has argued that this neglect is both a missed opportunity and a correctable one. Yet correcting it requires a real shift in mindset.

I vividly remember presenting the methodology for calculating sustainable development policy scores for the first time, showing that countries like Ghana and Senegal scored higher than many wealthy Western European nations. Afterwards, a portfolio manager approached me. He said he had enjoyed the talk but found the results unconvincing. In parts of Africa, he argued, people still 'live in huts' –so even if some progress had been made, how could investing there ever be considered sustainable?

His reaction was instinctive, but it revealed something important. Next to revealing limited awareness about life in developing countries, he was thinking about sustainable investing as the act of financing places that are already sustainable, rather than as a tool for advancing progress. He was also, without realizing it, conflating poverty with stagnation: assuming that because a country is relatively poor today, it must be failing. Ghana's 75% reduction in maternal mortality since 1985 is not the story of a country living in huts. It is the story of a country climbing the development ladder.

This is precisely the mindset shift the fox's approach demands. The first step is recognizing that the dominant tool for screening countries – sovereign ESG ratings – is structurally flawed as a guide to sustainable impact. As the World Bank has shown, these ratings reflect national income far more than policy quality. A hedgehog investor following that signal withholds capital from exactly the kind of progress the world needs more of. The fox asks a different question entirely: not where a country stands, but whether its policies are capable of driving it forward.

This chapter has shown that answering that question is possible – not perfectly, and not without data limitations, but well enough to build portfolios that direct capital toward countries advancing sustainable development. That said, the reach of sovereign bond investing has limits: 2.4 billion people live in countries that remain effectively invisible to mainstream bond investors, beyond the reach of capital allocation. Nevertheless, around 2.1 billion people live in countries that combine supportive sustainability policies with accessible bond markets – and for them, redirecting capital can make a tangible difference. Even modest reallocations from developed to emerging markets could expand the financing available for sustainable development, reduce borrowing costs for countries with credible sustainability plans, and unlock progress where it is needed most. Add engagement to the mix, and investors can do even more – using their voice to encourage governments to adopt more sustainable policies, and signalling that the global flow of capital has values as well as interests.

Sustainable investing in government bonds is not just possible; it is also urgent. And it can shift capital from where it is comfortable to where it is critical.

9

Epilogue

Scaling Sustainable Investing

Humanity stands at a crossroads. Our activities have become so profound that we have pushed the planet into a new geological epoch. Earth has left the relatively stable Holocene – during which human civilization flourished – and entered the Anthropocene, an era defined by lasting human influence on the climate and ecosystems [215]. We are driving climate change [161], have altered more than 75% of the Earth's land surface, disrupted two-thirds of the ocean, and are threatening around one million species with extinction [216]. These changes pose a real risk of destabilizing planetary health in ways that are neither linear nor predictable.

These are not abstract concerns. Climate extremes increasingly threaten lives and livelihoods. Nature loss undermines the ecosystems on which food, water, and economic stability depend [217]. Social challenges – from inequality and hunger to lack of access to basic services – remain widespread. And the geopolitical context is darkening. Armed conflicts, from Europe to the Middle East, are causing civilian suffering on a scale not seen in decades, erasing the development gains of years in a matter of months. The international legal and institutional frameworks built to prevent exactly these kinds of catastrophes are under severe strain. Meanwhile, geopolitical tensions are reshaping trade flows, energy markets, and the global financial architecture in ways that make long-term investment planning considerably more uncertain. Across all of these dimensions, the world is becoming harder to navigate – and the cost of ignoring sustainability, in its broadest sense, is rising.

J. A. van Zanten, *Big Money, Better World?*,
https://doi.org/10.1007/978-3-032-32442-9_9

Yet this is not the whole picture. Real sustainability transitions are taking place around the world. Renewable energy is accelerating at a pace that once seemed unimaginable [218]. New approaches to sustainable food systems are emerging [219]. Citizens around the world overwhelmingly care about climate change and nature loss [220].

Such shifts show that societies can move in the right direction and that change is possible, but also that it requires deliberate effort from every actor. Investors – by controlling how capital flows around the world – are one such actor. Not the most powerful, and not sufficient on their own. But not irrelevant either.

9.1 From Hedgehog to Fox

This book has argued that sustainable investing, done well, can make a meaningful difference. But done poorly – which is more common than the industry typically admits – it can mislead, misdirect, and ultimately fail the very goals it claims to serve. Understanding the difference is what this book has been about.

The dominant approaches to sustainable investing today behave like hedgehogs. They rally around one big idea – portfolio decarbonization, ESG integration, or impact monetization – and treat it as the answer to everything. The *specialist hedgehog* fixates on a single issue like climate change, ignoring the rest. The *filtering hedgehog* considers many issues but filters them all through one lens – like how ESG ratings view sustainability performance through the lens of financial materiality – leaving real-world impact out of the equation. And the *accountant hedgehog* reduces the plural complexity of sustainability to a single monetary value, creating an illusion of comparability that obscures more than it reveals. Each approach offers simplicity, structure, and a sense of control. However, each falls short by oversimplifying the messy reality of sustainable development.

The alternative is the fox. Unlike the hedgehog, the fox sees sustainability as it actually is: plural, interconnected, and full of trade-offs. A credible sustainable investing strategy must assess companies – and countries – across multiple sustainability dimensions simultaneously. It must deliver an overall judgement that prioritizes avoiding harm. And it must be grounded in sustainability science: not a static catalogue of findings, but a systems-based understanding of how environmental and social challenges interact, propagate, and compound. The fox's approach is harder to build and harder to market. Yet it is the only approach that reflects the complexity of the world it claims to address. As the physicist Richard Feynman warned: *'The first principle is that you must not fool yourself — and you are the easiest person to fool'.* In sustainable investing, where emotions run high and marketing often outpaces analysis, the fox resists this temptation. It grounds decisions in data, not ideology.

What the evidence shows is both humbling and encouraging. Sustainable investing is neither the dangerous placebo its harshest critics claim nor the world-saving force its most enthusiastic advocates promise. It can incentivize companies to improve, influence the cost of capital for sustainable and unsustainable activities, and shape the fields in which companies operate. Moreover, it can direct capital toward countries whose policies are advancing sustainable development. These are real effects. But they are modest and conditional – partly because investors exert influence indirectly, through asset allocation and active ownership, rather than by building solutions themselves, leaving them one step removed from the real-world outcomes they seek. The gap between what sustainable investing currently delivers and what it could deliver – if done rigorously and at scale – remains large. Closing that gap is what this book ultimately asks of its readers.

9.2 Why Scale Matters

Throughout this book, the evidence has pointed in a consistent direction: sustainable investing works better when it is bigger. The reason is that scale affects the signal sent by the two main levers of sustainable investing: exit and voice.

On exit – asset allocation the research shows that when enough investors screen consistently and redirect capital away from unsustainable firms toward sustainable ones, asset prices shift. That shift creates real incentives: management teams notice when their cost of capital rises, when their share price stagnates, and when sustainability-minded shareholders become harder to attract. But this mechanism only operates at scale. A handful of investors divesting from a coal company changes little – someone else buys. When a critical mass of investors moves in the same direction, using consistent criteria, the signal becomes harder to ignore. The same logic applies to the cost of capital for sustainable firms: the more the capital flows toward them, the cheaper it becomes for them to finance the solutions the world needs.

On voice – active ownership – scale matters just as directly. Engagement succeeds when investors have meaningful ownership stakes and when companies face credible consequences for non-compliance. Proxy voting carries more weight when it represents a substantial share of outstanding equity. Collaborative engagement initiatives like Climate Action 100+ derive their influence precisely from the financial weight behind them. A single investor asking a company to set science-based targets is easy to ignore, but 40 investors representing USD 17 trillion in assets are not.

The numbers illustrate how far current practice falls short of this potential. In 2023, private finance that harmed nature totalled USD 5 trillion, which is roughly 140 times more than the USD 35 billion flowing toward nature-positive outcomes [221]. Just 250 companies account for two-thirds of all negative impacts on nature, concentrated in food production, mining, and energy [222]. These imbalances will persist as long as sustainable investing remains a niche activity – as long as the capital doing real sustainability work is dwarfed by the capital that is not, and as long as sustainable funds remain satellite allocations at the edge of mainstream portfolios rather than competing at their core.

The fox views sustainable investing pragmatically: to matter at scale, sustainable investing must compete within the mainstream, not alongside it. A sustainable portfolio that cannot match the risk-return profile of an unrestricted one will always remain a satellite – something investors hold in addition to their core portfolio, not instead of it. The evidence in this book suggests that well-constructed, foxy sustainable portfolios can match conventional ones on diversification and factor exposure over the long run. Sustainable investors have a strong argument for building scale: the evidence suggests that returns need not be sacrificed.

Getting there requires overcoming several barriers that currently limit scale.

The first is the *divergence* of sustainability screens. When different investors use different methods to assess the same company and reach contradictory conclusions, the signal to management is noise rather than pressure. As research showed, even well-intentioned ESG ratings can diverge so sharply that the same company is simultaneously an ESG leader and an ESG laggard depending on who is doing the rating. A shared, science-based understanding of which companies are sustainable would make the collective signal far more powerful and far harder for companies to ignore.

The second barrier is *transparency*. Too many sustainability screens operate as black boxes, producing scores that investors cannot interrogate or explain. If asset owners cannot understand why a screen reaches a particular result, they cannot hold their managers accountable.

The third barrier is practical fit. Sustainability screens must be designed to work within core portfolio management, not as add-ons that create tracking error or factor distortions that make them unattractive to mainstream investors. The fox understands this pragmatism: the goal is not a purer but smaller pool of sustainable capital but a larger pool that moves markets. As Nina Seega of the Cambridge Institute for Sustainability Leadership argues, redirecting capital flows toward sustainable outcomes does not require complex new products or bespoke structures. The tools already exist: research, active ownership and asset allocation [223,224]. The question is not what to do, but who needs to act to make it happen at scale.

9.3 Time for Action

Scale does not emerge on its own. It requires deliberate choices by various actors. Two types of actors – investors and policymakers – are among the best positioned to unlock it.

For investors, the core implication of this book is straightforward: move from passive to active, and from hedgehog to fox. Passive investing has surged in recent years, and for good reasons – it is cheap, efficient, and difficult to beat. But sustainability is not passive [225]. It demands choices about which assets to hold, which risks to manage, and which impacts to prioritize. That means active decisions, and it means screening methods that are plural, science-based, and transparent enough for asset owners to understand and interrogate.

Asset owners in particular have a critical role to play. The principal-agent problem identified in Chapter 5 – where asset managers primarily pursue financial returns while asset owners increasingly want their capital to do more – will only be resolved if asset owners up their demands. That means asking hard questions of their managers: how does your sustainability screen work? What does it measure, and what does it miss? Does it capture financial or impact materiality? Does it reflect the plural, interconnected nature of sustainable development, or does it reduce everything to a single lens? These are important questions, and asking them is one of the most powerful things an asset owner can do to shift the industry toward foxier, more impactful approaches.

For policymakers, the most important single action for scaling sustainable investing is one that current regulation has largely avoided: pricing negative externalities. Companies generate harmful effects – polluting the environment, consuming scarce water, damaging human health, and emitting greenhouse gases – and in most cases do not pay the full cost of that harm. Citizens pick up the bill. Investors, in turn, do not systematically incorporate these costs into financial analysis because the market does not require them to. If companies had to pay for the damage they inflict, the financial calculus would shift dramatically. Unsustainable firms would become less profitable and less attractive as investments. Sustainable ones would benefit from a structural advantage that no amount of voluntary commitment currently provides. Pricing externalities would make sustainable investing more attractive while eroding the business case for unsustainable investing.

The second action concerns the regulation of sustainable finance itself. Here, the current approach has created a paradox. Sustainable investors face a maze of disclosure requirements, compliance costs, and reporting obligations. Unsustainable investors face none of these. If you invest in fossil fuels or cluster munitions, there is no extra reporting burden. But launch a sustainable fund, and extensive compliance follows immediately. This is, as the economist Arthur Pigou would have recognized immediately, exactly backwards.[1] Harmful activities should face greater scrutiny; beneficial ones should be encouraged. Regulators have the tools to correct this, and doing so would remove one of the most practical barriers to scaling sustainable investing within mainstream portfolios.

When investors and policymakers move together, the dynamics are reinforcing. Better disclosure gives foxy investors better data. Consistent screening gives companies clearer signals. Pricing externalities makes the financial case for sustainability self-evident rather than requiring investors to make it themselves. None of this happens automatically, but none of it is beyond reach either. The direction is clear, the tools exist, and the evidence is there. What remains is the will to act.

And institutional investors and policymakers are not the only actors with a role to play. Individual savers – whose pension contributions, savings accounts, and investment choices collectively represent an enormous pool of capital – can also make a difference, even if the levers available to them are more limited. The fox's framework applies here too, in simplified form: ask what your pension fund or fund manager is actually investing in, not just what it claims to stand for. Demand transparency on how sustainability screens work. And where choice exists between funds with different sustainability profiles, exercise it with the same scepticism this book has

[1] Arthur Pigou (1877–1959) was the Cambridge economist whose work gave rise to the term Pigouvian tax: the idea that negative externalities could be offset through taxation.

tried to encourage. Big money is, in the end, the aggregation of many individual decisions. That is both a humbling thought and an empowering one.

9.4 Final Thoughts

This book revolved around a metaphor that I was introduced to over candlelit dinners at Cambridge: the hedgehog and the fox. At the time, I was researching how companies can help work towards achieving sustainable development goals. The metaphor stuck. It offered a frame for analyzing the diverse and frequently conflicting impacts that companies have on the world – and it motivated me to embrace this plurality, searching for science-based approaches to evaluating how sustainable companies truly are.

In the years that followed, I designed sustainability screening methods for sustainable investors as a practitioner and studied these methods and their impacts as an academic. The metaphor stayed with me throughout: many sustainable investing strategies risked missing their point. Their hedgehog orientation led them to overlook the messy, interconnected reality of sustainable development. Better methods were needed – ones that resist the urge to reduce everything to a single vision. In short, sustainable investors require foxy sustainability methods.

Rethinking sustainable investing in this way can support real impact. Having examined the evidence, I find myself in a position that is neither cynical nor naive. Sustainable investing is not going to save the world on its own. But done well – with the fox's plurality, the scientist's rigour, and the investor's pragmatism – big money can contribute meaningfully to a better world. The evidence also suggests that investing more sustainably need not harm financial returns. What is more, sustainable investing helps prepare portfolios for a rapidly changing future, one in which ignoring sustainability is increasingly not an option.

Although a foxy approach to sustainable investing is more complex than the simpler hedgehog tactics that prevail today, it is also more exciting. It is honest about the complex – and intriguing – interactions that companies have with nature and society. Exploring these and embedding them into your portfolio is intellectually stimulating. It sets you up for some of the most rewarding conversations in finance: with asset owners who want their capital to reflect their values, with policymakers grappling with how markets can serve society, and with researchers pushing the boundaries of what we know. These conversations remind me regularly why this work matters.

On Illmatic, New York rapper Nas reminds the listener that '*The World Is Yours*'. The phrase captures something I believe – that we are not merely shaped by the world around us, but that we can shape it. This book has asked whether big money can build a better world. The answer, pursued with the fox's rigour and honesty, is yes. Not inevitably, not automatically, and not in ways that will save the planet on its own – but meaningfully and far more than the industry is currently delivering.

References

[1] UN Principles for Responsible Investment. (2026). *About the PRI*. Retrieved from https://www.unpri.org/about-PRI

[2] Global Sustainable Investment Alliance. (2021). *Global sustainable investment review*. Retrieved from http://www.gsi-alliance.org/wp-content/uploads/2021/08/GSIR-20201.pdf

[3] Global Sustainable Investment Alliance. (2023). *Global sustainable investment review*. Retrieved from https://www.gsi-alliance.org/wp-content/uploads/2023/12/GSIA-Report-2022.pdf

[4] NBIM. (2025). *About us*. Retrieved from https://www.nbim.no/en/about-us/norges-bank-investment-management/

[5] NBIM. (2025). *Climate and Nature Disclosures 2024*. Retrieved from https://www.nbim.no/contentassets/6fdfd333e6bf460f8e538b9b55a95bb7/gpfg-climate-and-nature-disclosures-2024.pdf

[6] PGGM. (2025). *Our investments*. Retrieved from https://www.pggm.nl/en/integrated-report/our-investments/

[7] Berlin, I. (2013). *The hedgehog and the fox: An essay on Tolstoy's view of history*. Princeton University Press.

[8] Krosinsky, C., & Robins, N. (Eds.). (2012). *Sustainable investing: The art of long-term performance*. Routledge.

[9] Renneboog, L., Ter Horst, J., & Zhang, C. (2008). Socially responsible investments: Institutional aspects, performance, and investor behavior. *Journal of Banking & Finance, 32*(9), 1723–1742.

[10] Meadows, D. H., Meadows, D. L., Randers, J., & Behrens III, W. W. (1972). *The limits to growth: A report for the Club of Rome's project on the predicament of mankind*. Potomac Associates.

[11] International Finance Corporation. (2004). *Who cares wins: Connecting financial markets to a changing world*. Washington DC: IFC.

J. A. van Zanten, *Big Money, Better World?*,
https://doi.org/10.1007/978-3-032-32442-9

[12] Galbreath, J. (2013). ESG in focus: The Australian evidence. *Journal of Business Ethics, 118*, 529–541.

[13] Jebe, R. (2019). The convergence of financial and ESG materiality: Taking sustainability mainstream. *American Business Law Journal, 56*(3), 645–702.

[14] Busch, T., Bruce-Clark, P., Derwall, J., Eccles, R., Hebb, T., Hoepner, A., ... & Weber, O. (2021). Impact investments: A call for (re)orientation. *SN Business & Economics, 1*(2), 33.

[15] Heeb, F., & Kölbel, J. (2020). *The investor's guide to impact.* Center for Sustainable Finance & Private Wealth, University of Zurich.

[16] Larcker, D. F., & Tayan, B. (2024). *Boeing 737 MAX.* Rock Center for Corporate Governance at Stanford University Working Paper (forthcoming).

[17] Robison, P. (2021). Boeing built an unsafe plane, and blamed the pilots when it crashed. *Bloomberg.* Retrieved from https://www.bloomberg.com/news/features/2021-11-16/are-boeing-planes-unsafe-pilots-blamed-for-corporate-errors-in-max-737-crash

[18] Cousté, G. (2024). After ESG: Is impact investment the new frontier for responsible investing? In *Sustainability stories: The power of narratives to understand global challenges* (pp. 263–267). Cham, Switzerland: Springer Nature Switzerland.

[19] Bolton, P., & Kacperczyk, M. (2021). Do investors care about carbon risk? *Journal of Financial Economics, 142*(2), 517–549.

[20] Garel, A., Romec, A., Sautner, Z., & Wagner, A. F. (2024). Do investors care about biodiversity? *Review of Finance, 28*(4), 1151–1186.

[21] Alves, R., De Waal, T., Hagens, M., ten Bosch, E., & van Dijk, M. A. (2025). *Thirsty for returns? The impact of water risk on the global stock market.* SSRN Working Paper.

[22] Enel. (2025). *Financial and environmental sustainability: The energy transition according to Enel.* Retrieved from https://www.enel.com/company/stories/articles/2025/05/modello-growth-environmental-financial-sustainability

[23] Enel. (2025). *Integrated Report 2024.* Retrieved from https://www.enel.com/investors/sustainability

[24] MarketScreener.com. (2025). *Enel SpA.* Retrieved from https://www.marketscreener.com/quote/stock/ENEL-S-P-A-70935/consensus/

[25] Edmans, A. (2023). The end of ESG. *Financial Management, 52*(1), 3–17.

[26] Edmans, A. (2024). Rational sustainability. *Journal of Applied Corporate Finance, 36*(2), 8–15.

[27] Heeb, F., Kölbel, J. F., Paetzold, F., & Zeisberger, S. (2023). Do investors care about impact? *The Review of Financial Studies, 36*(5), 1737–1787.

[28] European Commission. (2021). *Commission Delegated Regulation (EU) 2021/2139 of 3 June 2021 supplementing Regulation (EU) 2020/852 of the European Parliament and of the Council by establishing the technical screening criteria for determining the conditions under which an economic activity qualifies as contributing substantially to climate change mitigation or climate change adaptation and*

for determining whether that economic activity causes no significant harm to any of the other environmental objectives. Official Journal of the European Union, OJ L 442, 1–349.
[29] Top1000funds. (2025). *NBIM divests firms linked to Gaza and West Bank crisis.* Retrieved from https://www.top1000funds.com/2025/08/nbim-divests-firms-contributing-to-gaza-and-west-bank-crisis/
[30] Reuters. (2025). *Norway wealth fund excludes Caterpillar and five Israeli banks.* Retrieved from https://www.reuters.com/sustainability/society-equity/norway-wealth-fund-excludes-caterpillar-five-israeli-banks-2025-08-25/
[31] University of Cambridge. (2020). *Cambridge to divest from fossil fuels with 'net zero' plan.* Retrieved from https://www.cam.ac.uk/news/cambridge-to-divest-from-fossil-fuels-with-net-zero-plan
[32] Flammer, C. (2021). Corporate green bonds. *Journal of Financial Economics, 142*(2), 499–516.
[33] Beteta Vejarano, G., & Swinkels, L. (2023). *Social, sustainability, and sustainability-linked bonds.* SSRN Working Paper No. 4420618.
[34] O'Kelley, R., & Droste, A. (2021). *Why ExxonMobil's proxy contest loss is a wake-up call for all boards. Harvard Law School Forum on Corporate Governance.* Retrieved from https://corpgov.law.harvard.edu/2021/07/05/why-exxonmobils-proxy-contest-loss-is-a-wakeup-call-for-all-boards/
[35] Smyth, J., & Hancock, A. (2025, August 25). Exxon escalates attacks on EU's 'high-regulation' climate policies. *Financial Times.*
[36] *The Guardian.* (2026, April 23). *BP board suffers triple climate rebellion from shareholders.* Retrieved from https://www.theguardian.com/uk-news/2026/apr/23/bp-board-suffers-triple-climate-rebellion-from-shareholders
[37] Climate Action 100+. (2025). *Investors welcome RWE's climate target validation and coal phase-out progress.* Retrieved from https://www.climateaction100.org/news/investors-welcome-rwes-climate-target-validation-and-coal-phase-out-progress/
[38] Blitz, D., Swinkels, L., & van Zanten, J. A. (2021). Does sustainable investing deprive unsustainable firms of fresh capital? *The Journal of ESG and Impact Investing, 1*(3), 10–27.
[39] NBIM. (2025). *Decisions on exclusion.* Retrieved from https://www.nbim.no/en/news-and-insights/the-press/press-releases/2025/decisions-on-exclusion/
[40] Busch, T., Bauer, R., & Orlitzky, M. (2016). Sustainable development and financial markets: Old paths and new avenues. *Business & Society, 55*(3), 303–329.
[41] World Commission on Environment and Development. (1987). *Our common future.* Oxford University Press.
[42] GIIN. (2025). *What you need to know about impact investing.* Retrieved from https://thegiin.org/publication/post/about-impact-investing/
[43] Eccles, R. (2023). *'A whistleblower's tale' as told by Desiree Fixler. Forbes.* Retrieved from https://www.forbes.com/sites/bobeccles/2023/01/17/a-whistleblowers-tale-as-told-by-desiree-fixler/

[44] Reuters. (2025). *Deutsche Bank-owned asset manager DWS fined $27 million for greenwashing*. Retrieved from https://www.reuters.com/sustainability/german-asset-manager-dws-fined-25-mln-eur-greenwashing-case-2025-04-02/
[45] State Street. (2025). *Sustainable investing*. Retrieved from https://www.ssga.com/uk/en_gb/institutional/capabilities/esg
[46] Vanguard. (2025). *ESG investing*. Retrieved from https://investor.vanguard.com/investment-products/esg
[47] PIMCO. (2025). *Key takeaways from PIMCO's Sustainable Investing Report 2024*. Retrieved from https://www.pimco.com/nl/en/insights/key-takeaways-from-pimcos-sustainable-investing-report-2024
[48] Amundi. (2025). *Global Responsible Investment Policy*. Retrieved from https://www.amundi.com/institutional/files/nuxeo/dl/c44a7bb2-813b-4346-96e0-e3d695241d9b?inline=
[49] PRI. (2025). *What are the principles for responsible investment?* Retrieved from https://www.unpri.org/about-PRI/what-principles-for-responsible-investment
[50] IIGCC. (2025). *Our work*. Retrieved from https://www.iigcc.org/our-work
[51] UNEP FI. (2025). *UN-convened Net-Zero Asset Owner Alliance*. Retrieved from https://www.unepfi.org/net-zero-alliance/
[52] Finance for Biodiversity. (2025). *Our mission*. Retrieved from https://www.financeforbiodiversity.org/
[53] Nature Action 100. (2025). *Nature Action 100*. Retrieved from https://www.natureaction100.org/
[54] Van Zanten, J. A., & Rein, B. (2023). Who owns (un)sustainable companies? Examining institutional determinants of sustainable investing. *Journal of Cleaner Production, 422*, 138542.
[55] Galaz, V., Crona, B., Dauriach, A., Scholtens, B., & Steffen, W. (2018). Finance and the Earth system: Exploring the links between financial actors and non-linear changes in the climate system. *Global Environmental Change, 53*, 296–302.
[56] Gibson Brandon, R., Glossner, S., Krueger, P., Matos, P., & Steffen, T. (2022). Do responsible investors invest responsibly? *Review of Finance, 26*(6), 1389–1432.
[57] Nofsinger, J. R., & Varma, A. (2021). Keeping promises? Mutual funds' investment objectives and impact of carbon risk disclosures. *Journal of Business Ethics, 187*(3), 493–516.
[58] de Groot, W., de Koning, J., & van Winkel, S. (2021). *Sustainable voting behavior of asset managers: Do they walk the walk?* SSRN Working Paper No. 3783454.
[59] Kim, S., & Yoon, A. (2023). Analyzing active fund managers' commitment to ESG: Evidence from the United Nations Principles for Responsible Investment. *Management Science, 69*(2), 741–758.
[60] Trinks, A., Scholtens, B., Mulder, M., & Dam, L. (2018). Fossil fuel divestment and portfolio performance. *Ecological Economics, 146*, 740–748.

[61] Hong, H., & Kacperczyk, M. (2009). The price of sin: The effects of social norms on markets. *Journal of Financial Economics*, *93*(1), 15–36.
[62] Liang, H., Sun, L., & Teo, M. (2022). Responsible hedge funds. *Review of Finance*, *26*(6), 1585–1633.
[63] Sachs, J., Lafortune, G., Kroll, C., Fuller, G., & Woelm, F. (2022). *Sustainable Development Report 2022: From crisis to sustainable development: The SDGs as roadmap to 2030 and beyond*. Cambridge University Press.
[64] Follow the Money. (2022). *The Great Green Investigation*. Retrieved from https://www.ftm.eu/green-investments
[65] Van der Pol, L., & Woutersen, E. (2022). *Wat is de kleur van geld? Groen…? De Groene Amsterdammer*. Retrieved from https://www.groene.nl/artikel/wat-is-de-kleur-van-geld-groen
[66] Sanctuary, M., Lavenius, A., Parlato, G., Plue, J., & Crona, B. (2025). Green or brown: Are Article 8 & 9 fund portfolios different? *The European Journal of Finance*, *31*(15), 1948–1982.
[67] Scheitza, L., & Busch, T. (2024). SFDR Article 9: Is it all about impact? *Finance Research Letters*, *62*, 105179.
[68] Mongiovi, G. (2020). Was Keynes a socialist? *Catalyst: A Journal of Theory and Strategy*. Retrieved from https://catalyst-journal.com/2020/03/was-keynes-asocialist
[69] *The Economist*. (2022). *ESG should be boiled down to one simple measure: Emissions*. Retrieved from https://www.economist.com/leaders/2022/07/21/esg-should-be-boiled-down-to-one-simple-measure-emissions
[70] Morningstar. (2024). *Investing in times of climate change*. Retrieved from https://www.morningstar.com/sustainable-investing/investing-times-climate-change
[71] Responsible Investor. (2022). *MSCI makes implied temperature scores less precise after pushback*. Retrieved from https://www.responsible-investor.com/msci-makes-implied-temperature-scores-less-precise-after-pushback/
[72] Bolton, P., Reichelstein, S. J., Kacperczyk, M. T., Leuz, C., Ormazabal, G., & Schoenmaker, D. (2021). Mandatory corporate carbon disclosures and the path to net zero. *Management and Business Review*, *1*(3), 21–28.
[73] Coca-Cola HBC. (2025). *Sustainability Statement 2024*. Retrieved from https://www.coca-colahellenic.com/content/dam/cch/us/documents/oar2024/sustainabilitystatement-2024.pdf.downloadasset.pdf
[74] Heineken. (2025). *Annual Report 2024*. Retrieved from https://www.theheinekencompany.com/newsroom/download/06c1e830-3a72-45e1-8931-ab8c8388fc64/heineken-n-v-annual-report-2024-final-20feb2025.pdf
[75] European Commission. (2022). *EU taxonomy for sustainable activities*. Retrieved from https://ec.europa.eu/info/business-economy-euro/banking-and-finance/sustainable-finance/eu-taxonomy-sustainable-activities_en
[76] European Commission. (2025). *EU Taxonomy Navigator*. Retrieved from https://ec.europa.eu/sustainable-finance-taxonomy/activities/activity/273/view

[77] IPCC. (n.d.). *Figure SPM.7*. Retrieved from https://www.ipcc.ch/report/ar6/wg3/figures/summary-for-policymakers/figure-spm-7/

[78] Van Zanten, J. A., & Van Tulder, R. (2021). Improving companies' impacts on sustainable development: A nexus approach to the SDGs. *Business Strategy and the Environment, 30*(8), 3703–3720.

[79] S&P. (2025). *S&P 500 Biodiversity Index*. Retrieved from https://www.spglobal.com/spdji/en/indices/sustainability/sp-500-biodiversity-index/#data

[80] MSCI. (2025). *MSCI World Climate Paris Aligned Index*. Retrieved from https://www.msci.com/indexes/index/735619

[81] Amnesty International. (2025). *Meta's new content policies risk fuelling more mass violence and genocide*. Retrieved from https://www.amnesty.org/en/latest/news/2025/02/meta-new-policy-changes/

[82] Hartzmark, S. M., & Sussman, A. B. (2019). Do investors value sustainability? A natural experiment examining ranking and fund flows. *The Journal of Finance, 74*(6), 2789–2837.

[83] Pelizzon, L., Rzeznik, A., & Hanley, K. W. (2021). *The salience of ESG ratings for stock pricing: Evidence from (potentially) confused investors*. SAFE Working Paper, (No. 310).

[84] Dimson, E., Marsh, P., & Staunton, M. (2020). Divergent ESG ratings. *The Journal of Portfolio Management, 47*(1), 75–87.

[85] The Corporate Governance Institute. (2025). *Tesla's governance nightmare*. Retrieved from https://www.thecorporategovernanceinstitute.com/insights/news-analysis/teslas-governance-nightmare/

[86] Berg, F., Kölbel, J. F., & Rigobon, R. (2022). Aggregate confusion: The divergence of ESG ratings. *Review of Finance, 26*(6), 1315–1344.

[87] Rubinstein, R. (2025). *The Great ESG Con Job*. Retrieved from https://www.linkedin.com/pulse/great-esg-con-job-how-we-turned-saving-world-50-jerk-rubinstein-wwtoe/

[88] Suri, T., & Jack, W. (2016). The long-run poverty and gender impacts of mobile money. *Science, 354*(6317), 1288–1292.

[89] *The Guardian*. (2024). *Dutch airline KLM misled customers with vague green claims, court rules*. Retrieved from https://www.theguardian.com/world/2024/mar/20/dutch-airline-klm-misled-customers-green-claims-court-rules

[90] *Politico*. (2025). *Shein slapped with €1M greenwashing fine in Italy*. Retrieved from https://www.politico.eu/article/italy-fines-shein-greenwashing-fast-fashion-sustainability-emissions/

[91] ClientEarth. (2025). *Coca-Cola to revise 'misleading' recycling claims following legal complaint*. Retrieved from https://www.clientearth.org/latest/press-office/press-releases/coca-cola-to-revise-misleading-recycling-claims-following-legal-complaint/

[92] Tobacco Tactics, University of Bath. (2025). *British American Tobacco*. Retrieved from https://www.tobaccotactics.org/article/british-american-tobacco/

[93] U.S. Department of Justice. (2022). *Glencore entered guilty pleas to foreign bribery and market manipulation schemes.* Retrieved from https://www.justice.gov/archives/opa/pr/glencore-entered-guilty-pleas-foreign-bribery-and-market-manipulation-schemes

[94] SunPower. (2025). *About us.* Retrieved from https://us.sunpower.com/company

[95] Xylem. (2023). *Evoqua is now part of Xylem.* Retrieved from https://www.xylem.com/en-us/brand/evoqua/

[96] van Zanten, J. A. (2025). Measuring companies' environmental and social impacts: An analysis of ESG ratings and SDG scores. *Organization & Environment, 38*(3), 403–439.

[97] Bakker, P. (2013). Accountants will save the world. *Harvard Business Review.* Retrieved from https://hbr.org/2013/03/accountants-will-save-the-world

[98] Impact Economy Foundation. (2025). *Impact management and accounting with IWAF.* Retrieved from https://impacteconomyfoundation.org/impactweightedaccountsframework/

[99] Schoenmaker, D., & Schramade, W. (2023). Case study integrated valuation: Inditex. In *Corporate finance for long-term value* (pp. 289–321). Cham, Switzerland: Springer International Publishing.

[100] Shrimsley, R. (2018, May 9). *The value of parks—and why not everything should have a price.* Financial Times. Retrieved from https://www.ft.com/content/acfc7fc2-52e1-11e8-b3ee-41e0209208ec?syn-25a6b1a6=1

[101] van Zanten, J. A., & van Tulder, R. (2021). Towards nexus-based governance: Defining interactions between economic activities and Sustainable Development Goals (SDGs). *International Journal of Sustainable Development & World Ecology, 28*(3), 210–226.

[102] IPCC. (2022). Summary for policymakers. In P. R. Shukla, J. Skea, R. Slade, A. Al Khourdajie, R. van Diemen, D. McCollum, M. Pathak, S. Some, P. Vyas, R. Fradera, M. Belkacemi, A. Hasija, G. Lisboa, S. Luz, & J. Malley (Eds.), *Climate change 2022: Mitigation of climate change. Contribution of Working Group III to the Sixth Assessment Report of the Intergovernmental Panel on Climate Change.* Cambridge University Press. doi:10.1017/9781009157926.001

[103] Bansal, P., Durand, R., Kreutzer, M., Kunisch, S., & McGahan, A. M. (2025). Strategy can no longer ignore planetary boundaries: A call for tackling strategy's ecological fallacy. *Journal of Management Studies, 62*(2), 965–985.

[104] Whiteman, G., Walker, B., & Perego, P. (2013). Planetary boundaries: Ecological foundations for corporate sustainability. *Journal of Management Studies, 50*(2), 307–336.

[105] Williams, A., Perego, P., & Whiteman, G. (2025). Boundary conditions for organizations in the Anthropocene: A review of the planetary boundaries framework 10 years on. *Journal of Management Studies, 62*(4), 1811–1846.

[106] Crona, B., Peterson, G., Meacham, M., Parlato, G., Lade, S. J., Rocha, J. C., & Galaz, V. (2025). A systems approach to sustainable finance: Actors, influence mechanisms, and potentially virtuous cycles of sustainability. *iScience*, *28*(7), 1–18.

[107] Grewatsch, S., Kennedy, S., & Bansal, P. (2023). Tackling wicked problems in strategic management with systems thinking. *Strategic Organization*, *21*(3), 721–732.

[108] Dentoni, D., Roglic, M., Williams, A., & Bansal, P. (2026). A review of systems perspectives in sustainability: How systems properties convey systems-wide dynamics. *Journal of Management*, *52*(6), 2427–2468.

[109] Robeco. (2022). *Robeco's SDG Framework*. Retrieved from https://www.robeco.com/en-int/insights/2022/10/robecos-sdg-framework-from-niche-to-mainstream

[110] He, L., Lohre, H., & van Zanten, J. A. (2025). Sustainability matters: Company SDG scores need not have size, location, and ESG disclosure biases. *The Journal of Impact and ESG Investing*, *6*(1), 83–95.

[111] Bauckloh, T., Dobrick, J., Höck, A., Utz, S., & Wagner, M. (2024). In partnership for the goals? The level of agreement between SDG ratings. *Journal of Economic Behavior & Organization*, *217*, 664–678.

[112] Edmans, A., Gosling, T., & Jenter, D. (2025). *Sustainable investing in practice: Objectives, constraints, and limits to impact*. FEB-RN Research Paper No. 18/2024, HKU Jockey Club Enterprise Sustainability Global Research Institute - Archive, European Corporate Governance Institute – Finance Working Paper No. 1028/2024.

[113] Bauer, R., Ruof, T., & Smeets, P. (2021). Get real! Individuals prefer more sustainable investments. *The Review of Financial Studies*, *34*(8), 3976–4043.

[114] Hartzmark, S. M., & Shue, K. (2022). *Counterproductive sustainable investing: The impact elasticity of brown and green firms*. SSRN Working Paper No. 4359282.

[115] Neste. (2025). *Annual Report 2024*. Retrieved from https://www.neste.com/files/pdf/2taUTsTyORoQuJq9nfN3To-Neste_Annual_Report_2024_Review_BoD.pdf

[116] TotalEnergies. (2025). *More energy, less emissions: Sustainability & Climate 2025 Progress Report*. Retrieved from https://totalenergies.com/system/files/documents/totalenergies_sustainability-climate-2025-progress-report_2025_en.pdf

[117] Shell plc. (2025). *Annual Report and Accounts 2024*. Retrieved from https://www.shell.com/who-we-are/ourvalues/_jcr_content/root/main/section_874829968/list/list_item_copy.multi.stream/1752580693041/6c20b8111738b9a590ba145f0d1c4fa0e530dae0/shell-annualreport-2024.pdf

[118] Argente, D., Baslandze, S., Hanley, D., & Moreira, S. (2025). *Patents to products: Product innovation and firm dynamics* (NBER Working Paper No. 34592). National Bureau of Economic Research.

[119] Leippold, M., & Yu, T. (2026). Firm-level green innovation beyond patents. *Review of Finance, 30*(2), 423–458.

[120] UNFCCC. (2021). *New Financial Alliance for Net Zero Emissions launches.* Retrieved from https://unfccc.int/news/new-financial-alliance-for-net-zero-emissions-launches

[121] Fancy, T. (2021). *The secret diary of a 'sustainable investor'—Part 1.* Retrieved from https://medium.com/@sosofancy/the-secret-diary-of-a-sustainable-investor-part-1-70b6987fa139

[122] Dunn, K. (2021). *Does ESG investing actually make a difference?* Retrieved from https://fortune.com/2021/08/25/green-inc-tariq-fancy-esg-investing-pointless/

[123] Marti, E., Fuchs, M., DesJardine, M. R., Slager, R., & Gond, J. P. (2024). The impact of sustainable investing: A multidisciplinary review. *Journal of Management Studies, 61*(5), 2181–2211.

[124] Kölbel, J. F., Heeb, F., Paetzold, F., & Busch, T. (2020). Can sustainable investing save the world? Reviewing the mechanisms of investor impact. *Organization & Environment, 33*(4), 554–574.

[125] Rohleder, M., Wilkens, M., & Zink, J. (2022). The effects of mutual fund decarbonization on stock prices and carbon emissions. *Journal of Banking & Finance, 134*, 106352.

[126] Gantchev, N., Giannetti, M., & Li, R. (2022). Does money talk? Divestitures and corporate environmental and social policies. *Review of Finance, 26*, 1469–1508.

[127] Kölbel, J. F., & Lambillon, A. P. (2022). *Who pays for sustainability? An analysis of sustainability-linked bonds.* Swiss Finance Institute Research Paper No. 23.

[128] Solving for Climate. (2025). *Prof Alex Edmans: Does divestment from fossil fuels work?* Retrieved from https://podcasts.apple.com/gb/podcast/prof-alex-edmans-does-divestment-from-fossil-fuels-work/id1776500195?i=1000724487300

[129] Cheng, G., Jondeau, E., Mojon, B., & Vayanos, D. (2024). *The impact of green investors on stock prices* (NBER Working Paper No. 32317). National Bureau of Economic Research.

[130] Berk, J. B., & Van Binsbergen, J. H. (2025). The impact of impact investing. *Journal of Financial Economics, 164*, 103972.

[131] Gormsen, N. J., Huber, K., & Oh, S. (2024). *Climate capitalists* (NBER Working Paper No. 32933). National Bureau of Economic Research.

[132] Caramichael, J., & Rapp, A. C. (2024). The green corporate bond issuance premium. *Journal of Banking & Finance, 162*, 107126.

[133] Liang, H., Shih, Y. C., Sun, Y., Wang, Y., Wu, Z., & Yao, T. (2025). *Financing the green transition of "brown" sectors.* Singapore Management University School of Business Research Paper (forthcoming).

[134] Robeco. (2024). *Active Ownership Report: Q3*. Retrieved from https://www.robeco.com/files/docm/docu-robeco-public-engagement-report-2024-q3.pdf

[135] Heeb, F., & Kölbel, J. (2024). *The impact of climate engagement: A field experiment*. SAFE Working Paper No. 437, Leibniz Institute for Financial Research SAFE. https://doi.org/10.2139/ssrn.4711873

[136] Dimson, E., Karakaş, O., & Li, X. (2015). Active ownership. *The Review of Financial Studies, 28*(12), 3225–3268.

[137] Dimson, E., Karakaş, O., & Li, X. (forthcoming). Coordinated engagements. *Journal of Finance*.

[138] Marti, E., Fuchs, M., DesJardine, M., Slager, R., & Gond, J. P. (2023). A more impactful strategy for sustainable investing. *Harvard Business Review*. Retrieved from https://hbr.org/2023/09/a-more-impactful-strategy-for-sustainable-investing

[139] Heeb, F., Kölbel, J. F., Ramelli, S., & Vasileva, A. (2023). *Is sustainable finance a dangerous placebo?* Swiss Finance Institute Research Paper No. 23-46.

[140] Green Belt Movement. (2013). *GBM/Peponi House Charity Tree Planting Partnership and Environmental Conservation Initiative*. Retrieved from https://www.greenbeltmovement.org/node/485

[141] Soll, J. (2023). *Don't take financial advice from Ron DeSantis. Politico*. Retrieved from https://www.politico.com/news/magazine/2023/06/04/what-ben-franklin-could-teach-ron-desantis-00099483

[142] Florida Executive Office of the Governor. (2023). *Governor Ron DeSantis further prohibits woke ESG considerations from state investments*. Retrieved from https://www.flgov.com/eog/news/press/2023/governor-ron-desantis-further-prohibits-woke-esg-considerations-state-investments

[143] Alves, R., Krüger, P., & Van Dijk, M. (2025). Drawing up the bill: Are ESG ratings related to stock returns around the world? *Journal of Corporate Finance, 93*, 102768.

[144] Atz, U., Van Holt, T., Liu, Z. Z., & Bruno, C. C. (2023). Does sustainability generate better financial performance? Review, meta-analysis, and propositions. *Journal of Sustainable Finance & Investment, 13*(1), 802–825.

[145] Huij, J., Laurs, D., & van Zanten, J. A. (2025). The investment implications of sustainable investing. *Journal of International Money and Finance, 159*, 103372.

[146] Markowitz, H. M. (1990). *Nobel Prize lecture: Foundations of portfolio theory*. USA, Baruch College, The City University of New York, New York.

[147] Statman, M. (2017). *Finance for normal people: How investors and markets behave*. Oxford University Press.

[148] Bauer, R., Derwall, J., & Tissen, C. (2023). Private shareholder engagements on material ESG issues. *Financial Analysts Journal, 79*(4), 64–95.

[149] Barko, T., Cremers, M., & Renneboog, L. (2022). Shareholder engagement on environmental, social, and governance performance. *Journal of Business Ethics, 180*(2), 777–812.

[150] Sierra Club. (2025). *Investing in carbon bombs: How asset managers are using your savings to finance fossil fuel projects and fuel the climate crisis*. Retrieved from https://www.sierraclub.org/press-releases/2024/02/new-report-asset-managers-are-using-peoples-retirement-savings-finance

[151] Climate Safe Pensions. (2025). *Actors launch Retire Big Oil campaign, urging SAG-Producers Pension Plan to stop investing over $100M in fossil fuel companies*. Retrieved from https://climatesafepensions.org/retire-big-oil-launch-2025/

[152] New Private Markets. (2025). *PGGM wants measurable real world impact*. Retrieved from https://www.newprivatemarkets.com/pggm-wants-measurable-real-world-impact/

[153] Vasileva, A., van Zanten, J. A., & Swinkels, L. (2024). *Corporate sustainability and scandals*. SSRN Working Paper No. 4925213.

[154] Planetary Boundaries Science (PBScience). (2025). *Planetary Health Check 2025*. Potsdam Institute for Climate Impact Research (PIK), Potsdam, Germany.

[155] Crona, B., Folke, C., & Galaz, V. (2021). The Anthropocene reality of financial risk. *One Earth*, *4*(5), 618–628.

[156] Visaverge.com. (2025). *Extreme heat causes flight delays*. Retrieved from https://www.visaverge.com/news/extreme-heat-causes-flight-delays-for-southwest-delta-american-airlines/

[157] JPMorgan. (2022). *Climate change and the impact on agriculture*. Retrieved from https://www.jpmorgan.com/insights/global-research/sustainability/climate-change-agriculture

[158] World Economic Forum. (2025). *How climate hazards are reshaping business realities and responses*. Retrieved from https://www.weforum.org/stories/2025/03/how-climate-hazards-are-reshaping-business-realities-and-responses/

[159] Yoder, M. S. (2025). Impacts of drought on barge traffic and grain exports from the US, 2022 to 2023. *The International Trade Journal*, *39*(3), 250–267.

[160] *The Diplomat*. (2024). *How water scarcity threatens Taiwan's semiconductor industry*. Retrieved from https://thediplomat.com/2024/09/how-water-scarcity-threatens-taiwans-semiconductor-industry/

[161] IPCC. (2023). Summary for policymakers. In *Climate change 2023: Synthesis report. Contribution of Working Groups I, II and III to the Sixth Assessment Report of the Intergovernmental Panel on Climate Change* (Core Writing Team, H. Lee, & J. Romero, Eds., pp. 1–34). IPCC. doi:10.59327/IPCC/AR6-9789291691647.001

[162] Armstrong McKay, D. I., Staal, A., Abrams, J. F., Winkelmann, R., Sakschewski, B., Loriani, S., ... & Lenton, T. M. (2022). Exceeding 1.5°C global warming could trigger multiple climate tipping points. *Science*, *377*(6611), eabn7950.

[163] World Economic Forum. (2024). *Business on the edge: Building industry resilience to climate hazards.* Retrieved from https://reports.weforum.org/docs/WEF_Business_on_the_Edge_2024.pdf
[164] World Economic Forum. (2024). *Land degradation is a priority for business—Here's why.* Retrieved from https://www.weforum.org/stories/2024/12/tackling-land-degradation-business-priority/
[165] UNCTAD. (2025). *A rising tide, but growing risks: Ocean economy faces pressure and fragmentation.* Retrieved from https://unctad.org/news/rising-tide-growing-risks-ocean-economy-faces-pressure-and-fragmentation
[166] *The Guardian.* (2025). *Climate crisis on track to destroy capitalism, warns top insurer.* Retrieved from https://www.theguardian.com/environment/2025/apr/03/climate-crisis-on-track-to-destroy-capitalism-warns-allianz-insurer
[167] Klusak, P., Agarwala, M., Burke, M., Kraemer, M., & Mohaddes, K. (2023). Rising temperatures, falling ratings: The effect of climate change on sovereign creditworthiness. *Management Science, 69*(12), 7468–7491.
[168] United Nations. (2023). *The Sustainable Development Goals Report 2023: Special edition.* United Nations.
[169] Independent Group of Scientists appointed by the Secretary-General. (2023). *Global sustainable development report 2023: Times of crisis, times of change: Science for accelerating transformations to sustainable development.* United Nations.
[170] Biermann, F., Hickmann, T., Sénit, C. A., Beisheim, M., Bernstein, S., Chasek, P., Wicke, B., Kim, R. E., Kotzé, L. J., & Nilsson, M. (2022). Scientific evidence on the political impact of the Sustainable Development Goals. *Nature Sustainability, 5*(9), 795–800. doi:10.1038/s41893-022-00909-5
[171] van Zanten, J. A., & Putintseva, M. (2025). Evaluating governmental policies for the Sustainable Development Goals using hierarchical clustering. *International Journal of Sustainable Development & World Ecology, 32*(3), 322–340.
[172] Malekpour, S., Allen, C., Sagar, A., Scholz, I., Å P., Miranda, J. J., Al-Ghanim, K., Dube, O. P., Kanie, N., & Madise, N. (2023). What scientists need to do to accelerate progress on the SDGs. *Nature, 621*(7978), 250–254. doi:10.1038/d41586-023-02808-x
[173] ACEA. (2025). *New car registrations: -1.9% in H1 2025; battery-electric 15.6% market share.* Retrieved from https://www.acea.auto/pc-registrations/new-car-registrations-1-9-in-h1-2025-battery-electric-15-6-market-share/
[174] Volkswagen. (2025). *Annual Report 2024.* Retrieved from https://annualreport2024.volkswagen-group.com/_assets/downloads/entire-vw-ar24.pdf
[175] Mercedes-Benz Group. (2025). *Annual Report 2024.* Retrieved from https://group.mercedes-benz.com/documents/investors/reports/annual-report/mercedes-benz/mercedes-benz-annual-report-2024-incl-combined-management-report-mbg-ag.pdf

[176] Ford. (2025). *2024 Annual Report*. Retrieved from https://s205.q4cdn.com/882619693/files/doc_financials/2024/ar/Ford-2024-Annual-Report.pdf

[177] European Union. (2025). *Packaging waste*. Retrieved from https://environment.ec.europa.eu/topics/waste-and-recycling/packaging-waste_en

[178] U.S. Food and Drug Administration. (2025). *FDA's nutrition initiatives*. Retrieved from https://www.fda.gov/food/nutrition-food-labeling-and-critical-foods/fdas-nutrition-initiatives

[179] Holland & Knight. (2025). *Sustainable fashion law update: Critical legislation and compliance requirements for 2025*. Retrieved from https://www.hklaw.com/en/insights/publications/2025/04/sustainable-fashion-law-update-critical-legislation

[180] European Union. (2025). *Construction Products Regulation (CPR)*. Retrieved from https://single-market-economy.ec.europa.eu/sectors/construction/construction-products-regulation-cpr_en

[181] Caldecott, B., Clark, A., Koskelo, K., Mulholland, E., & Hickey, C. (2021). Stranded assets: Environmental drivers, societal challenges, and supervisory responses. *Annual Review of Environment and Resources*, *46*(1), 417–447.

[182] Sato, M., Gostlow, G., Higham, C., Setzer, J., & Venmans, F. (2024). Impacts of climate litigation on firm value. *Nature Sustainability*, *7*(11), 1461–1468.

[183] Callahan, C. W., & Mankin, J. S. (2025). Carbon majors and the scientific case for climate liability. *Nature*, *640*(8060), 893–901.

[184] Cowger, W., Willis, K. A., Bullock, S., Conlon, K., Emmanuel, J., Erdle, L. M., ... & Wang, M. (2024). Global producer responsibility for plastic pollution. *Science Advances*, *10*(17), eadj8275.

[185] Hansen, S. F., Bunde, C. T. H., Roy, M. A., Tickner, J. A., & Baun, A. (2024). Late lessons from early warnings on PFAS. *Nature Water*, *2*(12), 1157–1165.

[186] Schmidt National Law Group. (2025). *PFAS litigation: Current legal cases*. Retrieved from https://www.nationalinjuryadvocates.com/pfas-litigation-current-legal-cases/

[187] Reuters. (2023). *'Forever chemicals' were everywhere in 2023. Expect more litigation in 2024*. Retrieved from https://www.reuters.com/legal/litigation/forever-chemicals-were-everywhere-2023-expect-more-litigation-2024-2023-12-28/

[188] ChemSec. (2025). *'Just the start': The growing legal battle over PFAS in Europe*. Retrieved from https://chemsec.org/just-the-start-the-growing-legal-battle-over-pfas-in-europe/

[189] Krueger, P., Sautner, Z., & Starks, L. T. (2020). The importance of climate risks for institutional investors. *The Review of Financial Studies*, *33*(3), 1067–1111.

[190] Doeswijk, R. Q., & Swinkels, L. (2025). *The risk and reward of investing*. SSRN Working Paper No. 4937996.

[191] Gaspar, V., Garcia-Escribano, M., Prady, D., & Soto, M. (2019). *Fiscal policy and development: Human, social, and physical investments for the SDGs*. IMF Staff Discussion Note No. 19/003.

[192] Sachs, J., Sachs, L., Camelo Vega, A., & Willis, B. M. (2025). *Lowering the cost of capital for climate and SDG finance in emerging markets and developing economies (EMDEs)*. Columbia Center on Sustainable Investment.

[193] Barua, S. (2019). Financing Sustainable Development Goals: A review of challenges and mitigation strategies. *Business Strategy and Development, 3*(3), 277–293.

[194] UN DESA. (2025). *Ff4D 2025 outcome document*. Retrieved from https://financing.desa.un.org/ffd4/outcome

[195] Flammer, C., Giroux, T., & Heal, G. (2024). *Blended finance* (NBER Working Paper No. 32287). National Bureau of Economic Research.

[196] Tooze, A. (2025). *The end of development. Foreign Policy*. Retrieved from https://foreignpolicy.com/2025/09/08/adam-tooze-un-sustainable-development-goals-us-aid-finance-economy/

[197] Gratcheva, E. M., Emery, T., & Wang, D. (2020). *Demystifying sovereign ESG*. EFI Insight-Finance. World Bank.

[198] Gratcheva, E. M., Gurgy, B., Emery, T., Wang, D., Oganes, L., Linzie, J. K., Harvey, L., Marney, K., Murray, J., & Rink, R. (2021). *A new dawn: Rethinking sovereign ESG*. World Bank & J.P. Morgan.

[199] MacFeely, S. (2020). Measuring the Sustainable Development Goal indicators: An unprecedented statistical challenge. *Journal of Official Statistics, 36*(2), 361–378. doi:10.2478/jos-2020-0019

[200] Sachs, J., Schmidt-Traub, G., Mazzucato, M., Messner, D., Nakicenovic, N., & Rockström, J. (2019). Six transformations to achieve the Sustainable Development Goals. *Nature Sustainability, 2*(9), 805–814. doi:10.1038/s41893-019-0352-9

[201] Maestro, M., Pérez-Cayeiro, M. L., Chica-Ruiz, J. A., & Reyes, H. (2019). Marine protected areas in the 21st century: Current situation and trends. *Ocean & Coastal Management, 171*, 28–36.

[202] Edgar, G. J., Stuart-Smith, R. D., Willis, T. J., Kininmonth, S., Baker, S. C., Banks, S., ... & Thomson, R. J. (2014). Global conservation outcomes depend on marine protected areas with five key features. *Nature, 506*(7487), 216–220.

[203] Duarte, C. M., Agustí, S., Barbier, E., Britten, G. L., Castilla, J. C., Gattuso, J. P., ... & Worm, B. (2020). Rebuilding marine life. *Nature, 580*(7801), 39–51.

[204] Al-Abdulrazzak, D., & Trombulak, S. C. (2012). Classifying levels of protection in marine protected areas. *Marine Policy, 36*(3), 576–582.

[205] UNEP-WCMC. (2019). *User manual for the World Database on Protected Areas and World Database on Other Effective Area-Based Conservation Measures* (Version 1.6). UNEP-WCMC.

[206] National Geographic. (2025). *Ocean with David Attenborough shows bottom trawling as never before, underscores need for ocean protection.* Retrieved from https://news.nationalgeographic.org/ocean-with-david-attenborough-shows-bottom-trawling-as-never-before-underscores-need-for-ocean-protection/

[207] Sala, E., et al. (2022). *Declaration of support: Scientists call for the prohibition of all destructive fishing methods and industrial activities in marine protected areas.* Retrieved from https://static1.squarespace.com/static/5f241700a3b9e7503c42710c/t/631e89d3ccf13830735b5ccb/1662945747951/2022-09-11+-+Scientist+statement+-+EN.pdf

[208] Roberts, C., Béné, C., Bennett, N., et al. (2024). Rethinking sustainability of marine fisheries for a fast-changing planet. *npj Ocean Sustainability*, *3*, 41. doi:10.1038/s44183-024-00078-2

[209] New Economics Foundation. (2021). *Valuing the impact of a potential ban on bottom-contact fishing in EU marine protected areas.* Retrieved from https://seas-at-risk.org/wp-content/uploads/2021/05/Valuing-impacts-of-potential-ban-on-bottom-contact-fishing_NEF_-FINAL-for-publication.pdf

[210] Pauly, D., Zeller, D., & Palomares, M. L. D. (Eds.). (2020). *Sea Around Us: Concepts, design and data.* Retrieved from http://www.seaaroundus.org

[211] Greenwood, R., & Vayanos, D. (2010). Price pressure in the government bond market. *American Economic Review*, *100*(2), 585–590.

[212] Jansen, K. (2025). Long-term investors, demand shifts, and yields. *Review of Financial Studies*, *38*(1), 114–157.

[213] van Zanten, J. A., Sharma, B., & Christensen, M. (2023). Sustainability integration for sovereign debt investors: Engaging with countries on the SDGs. *Journal of Sustainable Finance & Investment*, *13*(3), 1300–1317.

[214] IPDD. (2022). *An investor-led sovereign engagement initiative that aims to halt deforestation.* Retrieved from https://weforum.ent.box.com/s/wetn6pwh063af0lpyvywu4qwx0to2u0d

[215] Lewis, S. L., & Maslin, M. A. (2015). Defining the Anthropocene. *Nature*, *519*(7542), 171–180.

[216] IPBES (2019). *Summary for policymakers of the global assessment report on biodiversity and ecosystem services of the Intergovernmental Science-Policy Platform on Biodiversity and Ecosystem Services.* S. Díaz, J. Settele, E. S. Brondízio E.S., H. T. Ngo, M. Guèze, J. Agard, A. Arneth, P. Balvanera, K. A. Brauman, S. H. M. Butchart, K. M. A. Chan, L. A. Garibaldi, K. Ichii, J. Liu, S. M. Subramanian, G. F. Midgley, P. Miloslavich, Z. Molnár, D. Obura, A. Pfaff, S. Polasky, A. Purvis, J. Razzaque, B. Reyers, R. Roy Chowdhury, Y. J. Shin, I. J. Visseren-Hamakers, K. J. Willis, and C. N. Zayas (eds.). IPBES secretariat, Bonn, Germany.

[217] Vogel, E., Donat, M. G., Alexander, L. V., Meinshausen, M., Ray, D. K., Karoly, D., ... & Frieler, K. (2019). The effects of climate extremes on global agricultural yields. *Environmental Research Letters*, *14*(5), 054010.